# Technocrats and the Politics of Drought and Development in Twentieth-Century Brazil

# Technocrats and the Politics of Drought and Development in Twentieth-Century Brazil

EVE E. BUCKLEY

The University of North Carolina Press  Chapel Hill

Set in Charis and Lato by Westchester Publishing Services
Manufactured in the United States of America

The University of North Carolina Press has been a member
of the Green Press Initiative since 2003.

Library of Congress Cataloging-in-Publication Data
Names: Buckley, Eve E., author.
Title: Technocrats and the politics of drought and development in
    twentieth-century Brazil / Eve E. Buckley.
Description: Chapel Hill : University of North Carolina Press, [2017] |
    Includes bibliographical references and index.
Identifiers: LCCN 2016054281 | ISBN 9781469634296 (cloth : alk. paper) |
    ISBN 9781469634302 (pbk : alk. paper) | ISBN 9781469634319 (ebook)
Subjects: LCSH: Brazil. Inspectoria Federal de Obras contra as Secas. |
    Drought relief—Political aspects—Brazil, Northeast. | Drought
    Relief—Social aspects—Brazil, Northeast. | Irrigation
    Engineering—Political aspects—Brazil, Northeast. | Irrigation
    Engineering—Social aspects—Brazil, Northeast. | Sustainable
    Development—Political aspects—Brazil, Northeast. | Sustainable
    Development—Social aspects—Brazil, Northeast.
Classification: LCC HD1696.B83 N6748 2017 |
    DDC 363.34/929098130904—dc23
    LC record available at https://lccn.loc.gov/2016054281

Cover illustrations: Top, small reservoir during drought, 1912 (Imagem IOC
[AC-E] 2-15.10, Acervo da Casa de Oswaldo Cruz, Dept. de Arquivo
e Documentação); bottom, federal rural health post, c. 1932 (Imagem BP
[F-VPP] 21-6, Acervo da Casa de Oswaldo Cruz, Dept. de Arquivo
e Documentação).

*To David*

*with love and thanks*

# Contents

Acknowledgments, xi

Abbreviations, xiii

A Note on Orthography, Currency, and Translation, xv

Introduction, 1
*Development Politics and Scientific Expertise*

1 Climate and Culture, 15
*Constructing Sertanejo Marginality in Modern Brazil*

2 Civilizing the Sertão, 46
*Public Health in Brazil's Hinterland, 1910s*

3 Engineering the Drought Zone, 82
*The Birth of IFOCS, 1909–1930*

4 Patronizing the Northeast, 125
*IFOCS under Vargas in the 1930s*

5 Watering Brazil's Desert, 147
*Agronomists and Sertão Reform, 1932–1955*

6 Modernizing a Region, 178
*Economists as Development Experts 1948–1964*

Conclusion, 214
*Science, Politics, and Social Reform*

Notes, 227

Bibliography, 249

Index, 271

# Illustrations, Maps, and Table

## Illustrations

Sertão landscape, 1912, 17

Starving sertanejos depicted in a Rio de Janeiro magazine, 1878, 25

Quixadá Reservoir, Ceará, 1912, 33

Small reservoir used by humans and animals during a drought, 1912, 64

Doctors and patients in a federal rural health post, Paraíba, 1923, 77

Drought agency reservoir under construction, 1912, 84

Workers at the São Gonçalo agricultural post, Paraíba, 1939, 128

Agronomist teaching a class on the use of a plow, Paraíba, 1939, 151

Home built by a worker at the São Gonçalo agricultural post, 1939, 160

Sertão home of a *coronel* and his family, 1912, 161

## Maps

The sertão drought zone in northeastern Brazil, 2

Sertão expeditions led by the Instituto Oswaldo Cruz, 1911–1913, 55

## Table

Capacity of reservoirs completed in Ceará, 1906–1962
and 1965–2006, 220

# Acknowledgments

This research was supported by National Science Foundation Grant No. 0217479 and by a National Endowment for the Humanities Summer Stipend, along with funding from the University of Delaware, its College of Arts and Sciences, and its Center for Global and Area Studies.

Over its long gestation, this project has incurred innumerable debts. I am grateful to the guidance of my mentors at the University of Pennsylvania—particularly Ann Farnsworth-Alvear, Steve Feierman, Rob Kohler, Susan Lindee, Janet Tighe, and the late Riki Kuklick. Colleagues Cari Constable, Erin McLeary, Lauren Nauta Minsky, Ian Petrie, Jeremy Vetter, Audra Wolfe, and many others provided moral support and good counsel at many stages. Profound thanks to Paulina Alberto, with whom I spent my initial months in Brazil.

At the University of Delaware, I have benefited from the mentorship and collegiality of Anne Boylan, Jesus Cruz, Rebecca Davis, Monica Domingues-Torres, Darryl Flaherty, Carla Guerrón-Montero, Carol Haber, John Hurt, Arwen Mohun, David Shearer, Patricia Sloane-White, and Owen White, among others. Particular thanks to my "write-on-site" *compañeras* Pascha Bueno-Hansen, Jennifer Gallo-Fox, Stephanie Kerschbaum, and Regina Wright.

In individual conversation and during conference panels I received valuable feedback on earlier versions of this work from Michael Adas, Chip Blake, Chris Boyer, Mark Carey, Oliver Dinius, Brodie Fischer, Seth Garfield, Thomas Klubock, John Krige, Stuart McCook, Gillian McGillivray, Eden Medina, Alfred Montero, Sara Phillips, Yovanna Pineda, Julia Rodriguez, Thomas Rogers, Martha Santos, Lise Sedrez, Nancy Stepan, Daryle Williams, and Mikael Wolfe. Special thanks to Lise, Marcos Chor Maio, and Tânia Salgado Pimenta for their hospitality in Rio de Janeiro.

In 2009, I was fortunate to meet Frederico de Castro Neves of the Universidade Federal do Ceará, who alerted me to the existence of drought works archives that had been indexed (in part) through the efforts of Almir Leal de Oliveira and his students. This led me to several productive months of research in Fortaleza, where I enjoyed the guidance and hospitality of

Aline Silva Lima and Anéssia Bayma, the librarian at the Departamento Nacional de Obras Contra as Secas (National Department for Works to Combat Droughts).

I was assisted in the final stages of manuscript preparation by my sister, Clare Buckley, and graduate student Mike Pospishil. Friend and colleague Gilberto Hochman of the Casa de Oswaldo Cruz in Rio de Janeiro smoothed the process of obtaining permission to include many of the photographs that appear in this book.

It has been a pleasure to work with Elaine Maisner and the staff at the University of North Carolina Press in preparing this manuscript for publication. My sincere thanks go to the press's two anonymous reviewers, who provided excellent suggestions to strengthen the book's argument and organization. I have done my best to adopt many of these; any errors that remain are my own.

I cannot credit my sons Casper, Henry, and Oliver with aiding the completion of this project, but they have made the last dozen years a joy and an adventure. I managed to complete this work in and around the demands of teaching and family life thanks to my husband, David Stockman, who ably looked after the home front during my extended research trips abroad and has been a constant source of moral and logistical support. My parents, Kate and Mike Buckley, also gamely pitched in at many points to keep things running smoothly, for which I am very thankful.

# Abbreviations

| | |
|---|---|
| AIB | Ação Integralista Brasileira (Brazilian Integralist Action) |
| BNB | Banco do Nordeste do Brasil (Bank of Northeast Brazil) |
| BNDE | Banco Nacional de Desenvolvimento Econômico (National Economic Development Bank) |
| CHESF | Companhia Hidro-Elétrica do São Francisco (São Francisco Hydroelectric Company) |
| CODENO | Conselho de Desenvolvimento do Nordeste (Economic Development Council for the Northeast) |
| CSC | Comissão de Serviços Complementares da Inspetoria de Secas (Commission of Services Complementary to the Inspectorate for Droughts) |
| CVSF | Comissão do Vale do São Francisco (São Francisco Valley Commission) |
| ECLA | UN Economic Commission for Latin America |
| ETENE | Escritório Técnico de Estudos Econômicos do Nordeste (Technical Office for Economic Studies of the Northeast) |
| DNOCS | Departamento Nacional de Obras Contra as Secas (National Department for Works to Combat Droughts) |
| GTDN | Grupo de Trabalho para o Desenvolvimento do Nordeste (Working Group for Northeast Development) |
| IFOCS | Inspetaria Federal de Obras Contra as Secas (Federal Inspectorate for Works to Combat Droughts) |
| IHB | International Health Board (of the Rockefeller Foundation) |
| IMF | International Monetary Fund |
| IOCS | Inspetaria de Obras Contra as Secas (Inspectorate for Works to Combat Droughts) |
| MVOP | minístro de viação e obras públicas (minister of transportation and public works) |

SUDENE    Superintendência de Desenvolvimento do Nordeste
(Superintendency for Northeast Development)

TVA        Tennessee Valley Authority

# A Note on Orthography, Currency, and Translation

I have adopted the spelling of names and institutions from historical sources when those are used consistently, even if they differ from contemporary Portuguese usage. Otherwise, I use standard contemporary Portuguese spelling, based on orthographic reforms of the mid-twentieth century.

Brazil's unit of currency from 1790–1942 was the old real (plural réis), written as Rs. $1. As its purchasing power declined over the nineteenth century, the more commonly used units of currency included the mil-réis, equivalent to 1,000 réis, and the conto, equivalent to one million réis. The mil-réis was written as Rs. 1$000 or sometimes 1$; the conto was written as Rs. 1:000$000 or sometimes 1:000$. In this book, the mil-réis and conto are the most commonly used units of currency. In 1933, US$1.00 was equal to Rs. 12$500 (12,500 réis, or 12.5 mil réis). By 1939 the value of the real had dropped significantly, and US$1.00 was equal to Rs. 22$500.

In 1942 the real was replaced by the cruzeiro, equal to Rs. 1$000 (one mil-réis) and written as Cr$1. Five new units of currency were introduced during the inflationary period between 1967 and 1993: the cruzeiro novo (NCr$) in 1967, cruzado (Cz$) in 1986, cruzado novo (NCz$) in 1989, cruzeiro (Cr$) again in 1990, and cruzeiro real (CR$) in 1993. Since 1994 the Brazilian currency has been a new real (plural reais), written R$1.00.

All translations herein are my own unless noted otherwise.

# Technocrats and the Politics of Drought and Development in Twentieth-Century Brazil

# Introduction
## Development Politics and Scientific Expertise
· · · · · · · · · · · · · · · · · · · · · · · · · · · · · · · · · · · · · · · · · · · · · · · · · ·

*Technocrats and the Politics of Drought and Development in Twentieth-Century Brazil* examines science and technology as vexed instruments of social reform in an impoverished Latin American region. It investigates twentieth-century Brazilian technocrats who saw themselves as offering a middle road between the reactionary conservatism of landowning elites and the revolutionary impulses of leftist reformers. Central to this study is climate unpredictability and the risks it presents in landscapes of entrenched poverty. This is a topic with increasing resonance as communities around the globe confront more extreme fluctuations in weather patterns. Ultimately, this book asks to what extent scientific expertise can solve pressing social problems—particularly, glaring inequities in wealth and security. It highlights the constraints on technocrats as agents of social change.

The regional focus is Brazil's semiarid northeastern hinterland known as the *sertão*. In Brazil's colonial era, cattle ranches expanded from the sugar-exporting coast (the economic and political center of Portugal's flourishing colony) to provide meat and muscle power for plantations. The sertão's fortunes as a ranching and agricultural economy waxed and waned in response to global competition and demand. By the beginning of the twentieth century, elites in Brazil's more dynamic south viewed the sertão as chronically backward, plagued by a feudal landowning structure and the perceived deficiencies of its mixed-race population. Yet national leaders also saw renovating the sertão as essential to their modernizing ambitions. Among the persistent challenges to which the sertão has been subject is periodic drought. The establishment of an agency to combat drought in 1909 launched the federal government's most significant investment in the region. Over the twentieth century, the Departamento Nacional de Obras Contra as Secas (DNOCS; National Department for Works to Combat Droughts) conducted geographic surveys, constructed road networks, built thousands of reservoirs of varying sizes, and planned irrigation systems that were intended to form the nuclei of smallholder

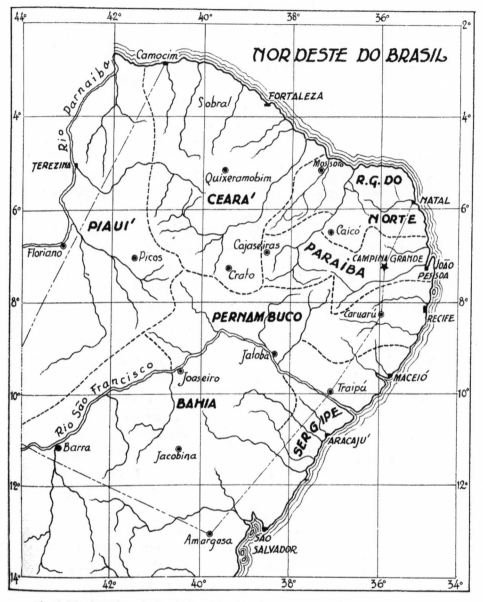

The sertão drought zone in northeastern Brazil. Source: *Boletim da Inspetoria Federal de Obras Contra as Secas*, 10, no. 1 (1938): n.p.

agricultural colonies. All of these projects aimed to mitigate vulnerability to drought and reduce the suffering of migrants who streamed toward coastal capitals during each calamity in search of food, medical care, and employment.

In the twentieth century, the ascendancy of professions such as civil engineering and agronomy within Brazil's drought agency mirrored the rise and fall of those professions in development agencies around the world. Brazil's drought technocrats modeled their efforts explicitly on development programs in impoverished U.S. regions, including southern hookworm eradication by the Rockefeller Foundation, dam construction by the U.S. Bureau of Reclamation, and the multi-faceted regional planning apparatus of the Tennessee Valley Authority. Like many of their global peers, DNOCS staff believed they could alter the social landscape of the sertão through infrastructural and hygienic engineering. However, northeastern Brazil's landowning elite as well as its laboring class often rejected scientists' embrace of new technologies and work regimes. The conflicting agendas of landowners, federal bureaucrats, and agricultural workers in the sertão help to explain why drought, famine, and poverty remained searing crises throughout the twentieth century. They illustrate the frequent tensions between a universalizing technocratic vision of progress and the particular cultural and political contexts in which scientific reformers operated. DNOCS personnel managing sertão construction sites often advocated on behalf of their impoverished manual laborers (in opposition to their superiors and regional elites) while simultaneously battling what they perceived as the ignorance and recalcitrance of people whom they strove to help.

DNOCS's archival records and publications, examined for the present study in Rio de Janeiro (the national capital from the late eighteenth century until 1959) and in Recife and Fortaleza (two northeastern state capitals), indicate differences in the reformist ideology of a range of Brazilian development professionals over the twentieth century. They reveal disagreements among drought agency personnel who interacted with famished sertão inhabitants at varying degrees of proximity, and they illuminate shifts in the local and global politics of technocratic reformism. Fundamental tensions between sertão social groups ultimately undermined drought technocrats' cherished goal of reaching rational, scientific solutions to what were inextricably political problems of inequality and poverty.

By focusing on the cohorts of technical experts who staffed and oversaw DNOCS and related federal agencies from 1909 through the 1960s and emphasizing the ideologies and reformist visions that underlay their efforts, this book analyzes the varied politics of public health sanitarians, agronomists, development economists, and others who hoped to transform the sertão by targeting climatic instability. Brazilian scholars who have analyzed DNOCS, particularly political scientists, often depict the agency's work as contributing to a "drought industry" that funneled federal funds to northeastern politicians and their landowning clients without addressing the persistent vulnerabilities of the landless poor. It seems indisputable that DNOCS's twentieth-century projects increased the security and power of regional landowners without markedly improving the fortunes of those most affected by drought. Yet close examination of DNOCS's project records reveals a more complex and nuanced story of agency technocrats' ambitions and accomplishments. In their interactions with politicians, sertão inhabitants, and government officials, many of DNOCS's civil engineers and agronomists exhibited genuine sympathy for the migrants whom they employed at agency worksites during droughts and a pronounced distaste for Brazil's chronic inequalities. Like other development professionals in the twentieth century, they were determined to improve the lot of those who lived in the sertão—the *sertanejos*—in keeping with their own middle-class views of progress and modernization—through rational, scientific management of the landscape and economy. Yet many participants in this drama ultimately concluded that there was no route to this end absent a starkly political struggle over access to water and land.

*Technocrats and the Politics of Drought and Development in Twentieth-Century Brazil* thus examines an influential segment of modern Latin American society that hoped to navigate a middle path between entrenched conservatism and social rupture—what historian Michael Ervin refers to in another Latin American context as a "middle politics."[1] Across the twentieth century, Latin American technocrats occupied bureaucratic positions in governmental and international organizations that aimed to reduce poverty using the tools of science. Brazil's politicians and social reformers hoped that technical experts could cure the sertão's chronic ills without encouraging broadly Marxist movements for social change that sometimes threatened to upend the social order (particularly after 1950). As archival records reveal, the drought agency's professionals worked at an uncomfortable intersection where technology's tantalizing promise of an apolitical means to end poverty collided with the stark probability that only open confrontation with those who monopolized land and water could reduce the dependence and vulnerability of the poor. DNOCS administrators' often naive confidence in technical solutions to long-standing injustices was endemic in twentieth-century Latin America, despite the repeated failure of similar endeavors to reduce inequities that had persisted since the colonial period. Their challenges highlight the limited menu of politically viable options that confronted middle-class Latin American professionals who hoped to guide their nations toward a social progress modeled in many ways (sometimes ill-advisedly) on the experience of the United States.

· · · · · ·

This project was inspired by studies of development failure by economist Amartya Sen, anthropologists James Ferguson and Arturo Escobar, historian James Scott, and political theorist Timothy Mitchell.[2] Several decades ago Sen posited that chronic hunger is a political issue rather than a technical one. His pivotal insight that famines result from crises of *access* to food rather than from insufficient food *supply*, undergirds many subsequent analyses—including mine. Ferguson's analysis of the "antipolitics" of development projects in Lesotho encouraged scholars to examine what such endeavors *do* accomplish, even (or especially) when they do not achieve the transformative outcomes that are their purported raisons d'être. One of his central insights is that rural development efforts restructure power, often increasing state authority over hinterland regions even as the agencies overseeing the projects vigorously deny having any political agenda. Focusing on Latin America, Escobar argues that modernization projects have consistently benefited the "social technicians" who oversee them more than the recipients of such aid, who can be harmed by outsider interference in local production and governance. Scott's widely influential book *Seeing Like a State* criticizes a twentieth-century ideology that he calls "high modernist" faith in technocratic expertise. This legitimated aggressive, state-sponsored social engineering, sometimes with disastrous results. Similarly, in *Rule of Experts* Timothy Mitchell highlights the perils of hubristic technocratic planning. Given the impossibility of understanding or accurately assessing the myriad variables at play in any culture or economy, the results of ambitious development schemes are unpredictable and the consequences for targeted populations can be dire. Like Ferguson, Mitchell deconstructs development rhetoric to reveal the unstated political objectives that are masked by framing technocratic intervention as a means of aiding the very poor.

This book is also rooted in the social history of science—particularly in studies of science and public health in colonial and postcolonial Africa and India. It is indebted to pathbreaking histories of Latin American science published by Marcos Cueto, Nancy Stepan, Stuart McCook, Julia Rodriguez and Eden Medina,[3] who analyze the role of scientific personnel in envisioning and pursuing modernization projects within their respective nations. Taken together, their work illustrates both the fundamentally political nature of technocratic ambitions and the range of political ideologies that Latin American sanitarians, anthropologists, agronomists, and others have embraced. My focus on drought as a central problem also enables engagement with the field of environmental history, as pioneered in Brazil by Warren Dean. Recent books that relate these subfields include Mikael Wolfe's study of agronomists and land reform in revolutionary Mexico and Mark Carey's analysis of the intersection between ecological and social vulnerability in the glacial ranges of the Peruvian Andes.[4] More closely linked to this book's regional focus is Thomas Rogers's interpretation of changes in ideological and material relationships to land among participants in the sugar industry that has long dominated northeastern Brazil's coastal region.[5]

Influenced by this body of research, the present volume focuses on several cohorts of twentieth-century Brazilian technocrats, attentive to both their changing global influences and adaptations to local realities. It adopts the model deployed by Michael Ervin to trace the "middle politics" of Mexican agronomists in the 1930s who, as agents of the revolutionary state, negotiated with those above and below them in the class hierarchy. The Brazilian sanitarians, civil engineers, agronomists, and economists who devoted their careers to mitigating drought's impact on the sertanejo poor believed firmly in the capacity of their particular expertise to address such problems; they were not, as some analyses of Brazil's "drought industry" suggest, uncaring instruments of an industrializing state. Some were deeply moved by the plight of the desperate people they encountered, and disturbed by the intractability of the recurring crisis. But these men were misled by their belief that one could arrive at rational, scientifically grounded solutions to what was fundamentally a political issue of unequal control over food and water. As with so many challenges in economic development, the central problem was that technocratic solutions adequate to end the famine and dislocation precipitated by droughts were not politically viable. Such reforms would have entailed a degree of social reconfiguration that incited vigorous obstruction by landowning elites—and that were, in many cases, beyond the imaginations of the technical personnel employed by the federal drought agency. As Tania Li has argued in another context, these men were trained to render complex problems technical, which often meant ignoring inconvenient political dimensions.[6] As members of a growing professional middle class in a self-consciously modernizing nation, they were rarely inclined to pursue measures that might radically upend a social order in which they enjoyed an increasingly secure position.

## The Setting

Stories of drought and migration in northeastern Brazil's backland sertão are legion across the twentieth century. Novels, films, newspaper accounts, and popular songs depict stoic sertanejos in exodus from their parched and searing homeland, trudging desperately toward coastal oases in search of food and temporary employment. Those who survive the arduous journey gather in makeshift refugee camps, where disease runs rampant and many die. Even in recent years, drought episodes have provoked the tragic but predictable migration of thousands of famished sertanejos fleeing their homes in hope of finding aid.

Descriptions of the misery caused by drought in the sertão have been reported in the Brazilian media since at least the 1870s. Major droughts have occurred at irregular intervals every couple of decades since, and sometimes more frequently. In response to pleas for assistance from regional politicians, Brazil's federal government established a series of agencies during the twentieth century with a mandate to mitigate the drought problem and its accompanying humanitarian crisis. In response to the notoriously brutal 1915 drought, urban governments in northeastern Brazil organized encampments to control amassed migrants from the sertão. Disease thrived within these camps, however, and at times cadavers were deposited along the fences to await collection by the authorities. As an alternative to this macabre and threatening situation, the federal drought agency determined by 1919 (under Aarão Reis's direction) that it should "secure the man in the field," providing aid within the sertão itself. New hydrologic and transportation infrastructure would be constructed in the sertão by male heads of migrant families, in return for meager rations of beans and raw cane sugar to feed their starving families. When the drought agency's engineers and agronomists traveled into the interior to oversee construction at these "work fronts," they encountered poor sertanejos, often for the first time. One engineer who oversaw such a worksite during the 1980s told me that some of her employees worked an extra six-hour shift in the brutal heat just to receive the modest snack provided at the end. It was, she remarked sadly, "no way to treat a dog."

The Nordeste (Northeast) was colonial Brazil's wealthiest region throughout centuries of slave-based cane sugar production. Its global dominance in this export industry faded by the late nineteenth century, but the social structure of landholding families who dominate political life as the patrons of a landless, tenant farming class persisted. The twentieth century's poorest sertanejos, descendants of native *índios*, escaped slaves, and others who sought autonomy in the unforgiving northeastern backlands, lived on the periphery of an economically decadent region. Many national and regional elites saw them as racially and culturally unsuited to modernization. South Brazilian exporters and industrialists, whose interests drove national policies, largely overlooked the problems of the Nordeste. When slavery was finally abolished in the late 1880s, southern coffee growers obtained state subsidies for Italian and German immigrants rather than hire their free northeastern compatriots of dubious racial heritage.

## Drought and Development

Landless sertanejos were vulnerable to drought and famine because of political dynamics governing the sertão and Brazil as a whole during the twentieth century. To elaborate upon a phrase coined by Mike Davis, they were subject to quadruple peripheralization: "the underdevelopment of Brazil within a world financial system dominated by British (and later U.S.) capital; the declining economic and political position of the Northeast region in relation to São Paulo; . . . the sertão's marginality within state politics controlled by coastal plantation elites," and the dependence of landless sharecroppers and ranch hands on their patron employers.[7] Droughts and the suffering they caused starkly revealed poor sertanejos' marginality.

For northeastern Brazil's landowning and political elite, climate instability was a convenient scapegoat for what many critics of the region's "drought industry" viewed as problems of political economy. Landless tenant farmers and ranch hands in the sertão were vulnerable to famine and disease for multiple reasons; recurrent droughts were one contributor to their vulnerability, but there were myriad political and social agents as well. The civil engineers and agronomists sent to oversee public works in the drought zone, and the economists charged with drafting regional development plans by the 1950s, identified a range of factors that made drought a crisis for poor rural families. As the region's most prominent economic planner, Celso Furtado, would argue by 1960, humanitarian crises spurred by drought were merely a symptom of endemic ills, particularly the concentration of land and wealth in a few hands and the imbalances in political power that stemmed from this.

During its first half century, DNOCS, which was established as an inspectorate in 1909, adopted a variety of approaches to reducing the crises precipitated by drought in the sertão. These approaches broadly mirrored the dominant trends in regional economic development embraced elsewhere in the world during those same decades. As droughts recurred, a sequence of professions (civil engineering, public health, agronomy, and economic planning) were given authority within DNOCS in hopes that their paradigms would remedy the shortcomings of previous strategies for drought amelioration. Each of these professions had also become influential in regional development agencies elsewhere, and staff from related U.S. agencies (such as the Bureau of Reclamation and the Tennessee Valley Authority) served as short-term advisers on new DNOCS projects. But nationalist critics of these foreign models feared that Brazil's development experts placed undue faith in technological infrastructure (dams, roads, and irrigation systems) without attending to the social dynamics that influenced the sertão's population and economy.

Civil engineers sent to oversee federal projects funded as relief efforts during droughts often believed that the sertanejo workers in their care must be acculturated to a more progressive mind-set and become less resigned to the whims of nature. The engineers' reformist crusade in the northeastern hinterland targeted both the corruption of local elites (such as *fornecedores*, who sold essential goods at exorbitant prices under conditions of acute scarcity) and the recalcitrance of farmers who did not strive to improve their families' economic circumstances through more efficient and rationalized labor regimes. Similarly, agronomists' ambitions for the sertão foundered on the shoals of sertanejos' distaste for embracing a more intensive cultivation regime, as well as on the more predictable elite resistance to fundamental social change. As one agronomist ruefully remarked, young sertanejo men preferred to dwell on their lovers, parties, or friends rather than on efficient work.[8]

Engineers and agronomists working in the sertão during the 1910s–50s saw themselves as rational middle-class professionals who could mediate between the self-interest of landowning elites and the desperation of humble sertanejo cultivators. They hoped to reduce elites' monopoly over natural resources and wealth without provoking violent confrontation and to tutor newly established smallholders in more disciplined production for their own and the nation's benefit. All this was to be accomplished through improved technology—particularly dams, roads, and irrigation canals. Upon arrival in the scorched northeastern interior, surrounded by thousands of nearly starving workers and their families, Brazil's drought technocrats quickly realized how much social reorganization and acculturation to their own values would be required in order to achieve such transformation. Irrigated smallholding, in particular, comprised both a technical venture (irrigated farming) and a politically radical step (providing land to the disenfranchised). But the agronomists and their engineering colleagues were tasked with effecting social change indirectly, relying on technology to circumvent political confrontation.

Although previous scholars' overarching depiction of a drought industry that fundamentally served the interests of landowning elites is accurate, this outcome did not arise due to drought technocrats' ignorance of, complicity in, or disregard for power imbalances in the sertão. The drought agency's technological infrastructure failed to end the suffering of the sertanejo poor over many decades because the social dynamics that made drought a recurrent humanitarian crisis were more complex than the agency's technocrats initially understood. The men sent to solve the problem of drought in Brazil's semiarid interior were not equipped by training, political temperament, or bureaucratic position to confront regional power brokers in the ways that genuine sertão transformation would have required. Beginning in the First Republic, the Brazilian state employed civil engineers and agronomists as apolitical agents of change, hoping that the scientists could increase the wealth and security of the sertanejo poor while avoiding conflict. This limited conception of the technocrats' role enabled the region's dominant political actors (large landowners) to get what they most wanted from the federal agency—namely, infrastructural improvement on their own land.

## Historical Overview

The time period covered by this study opens with the Great Drought of 1877–79, an episode that brought the sertão's recurrent calamity to national attention. This allowed northeastern elites to demand improvements to their region on behalf of poor sertanejos. In 1897 a protracted standoff between sertanejo followers of a millenarian prophet and the Brazilian military awakened southern politicians and intellectuals to the contrast between their hopes for a modern republic and the dismal conditions of their compatriots in the rural northeast. This impression was reinforced a decade later when medical personnel from the Instituto Oswaldo Cruz toured the sertão and reported on the overwhelming disease burden borne by its population. Such reminders of the cultural and physical differences separating urban elites from Brazil's poorest citizens mocked the modernizing vision evident in the sanitary and architectural reforms undertaken in Rio de Janeiro at the turn of the twentieth century.

The federalist First Republic, established in 1889, was a period of increasing regional differentiation during which political power lay primarily with state governors and the party machines that supported them. The economically decadent Nordeste region was losing wealth and influence relative to the coffee-exporting southern states. Physician Belisário Penna, the first director of "sanitization and rural prophylaxis" in Brazil's Department of Public Health, wrote forcefully during the 1910s and 1920s about the need to strengthen rural workers through modern hygiene and medicine, particularly in the northeast. He campaigned for numerous reforms, including greater provision of medical care and preventive health measures in rural areas and land redistribution to improve the political and economic security of the poor. Penna emphasized the need for renewed social investment in order to achieve the vision of national progress promised by republican modernizers.

During the decades when Penna was most active, engineers proposed a series of infrastructural improvements for the sertão. Their positivist education gave them confidence in the potential of rational planning to achieve social progress. At the turn of the century, engineers were involved in extending railroad and telegraph lines to enable more rapid communication between Brazil's coast and its expansive interior. They found further employment in the Inspetaria de Obras Contra as Secas (IOCS; Inspectorate for Works to Combat Droughts), established in 1909 and funded more generously during the 1920s and 1930s. Along with dam and reservoir construction, IOCS undertook rail line and road extension in the northeast, sponsored climatic, geologic, and botanical studies of the region, and funded rural health posts. IOCS's work followed models set by the Rockefeller Foundation's disease eradication programs in the southern United States and the U.S. Bureau of Reclamation's irrigation projects in western states, as well as technical development efforts undertaken in British colonial India and elsewhere.

Brazil's First Republic ended with a bloodless revolution in 1930 led by middle-class reformers and elites who were dissatisfied with the dominance of Sao Paulo's coffee oligarchs over national policy. The coup landed Getúlio

Vargas in the presidential palace, and he remained there until 1954, though with a hiatus from 1945 to 1951. Vargas was committed to a program of national industrialization under a strong central government, and the sertão was symbolically important to his administration as an impoverished but historically significant region that held great potential for modernization. Particularly in the 1930s, sertanejos were depicted in nationalist discourse as possessing both positive and negative racial characteristics. Because their ancestors included early Portuguese colonists, escaped slaves from the coastal sugar plantations, and native *índios,* Vargas often upheld sertanejos as the quintessential Brazilians—and thus the most appropriate subjects of his cultural and economic modernization efforts. Yet critics of federal aid to the sertão worried that its inhabitants were hopelessly backward, condemned to a primitive state by their mixed racial heritage and long adaptation to a harsh climate. And despite Vargas's rhetorical celebration of sertanejos' importance to modern Brazil, he committed significant resources to the sertão only during droughts that occurred at the beginning and end of his long administration. For most of his two decades in office, Vargas's attention was absorbed by the industrialization of coastal capitals and the demands of urban workers who comprised his political base. The drought agency expanded its agenda during these years to commence irrigation projects in the sertão, but the tangible results of this new orientation were minimal.

The president who followed Vargas, Juscelino Kubitschek, pursued an ambitious national development agenda during the late 1950s, focused primarily on south-central Brazilian industry and transportation infrastructure. Kubitschek acknowledged the need for greater federal investment in rural areas, including the sertão; he assembled advisory committees on northeastern development that comprised politicians and federal bureaucrats familiar with the region. These were led by economist Celso Furtado, a Nordestino who had worked for several years with Argentine economist Raul Prebisch at the UN's Economic Commission for Latin America in Chile. Furtado offered plans for northeastern development that aimed to diversify the sertão's economy and integrate rural and urban production. In 1959 he helped to draft a modest proposal for land reform that would convert underutilized sertão properties into irrigated smallholder settlements. Many politicians associated this effort with a more radical agrarian reform movement that emerged during the 1950s on behalf of agricultural workers known as the Peasant Leagues. Conservative northeastern elites feared Furtado's growing political influence. Their concerns about the revolutionary potential of his development plans helped to spawn a military coup that ousted left-leaning President João Goulart in 1964. Furtado went into exile, and many of his proposals were eliminated from the agendas of federal agencies working in the northeast during the decades that followed.

· · · · · ·

Chapter 1 establishes the historical and sociological reasons for sertanejos' marginality in twentieth-century Brazil. It considers how those with greater political power perceived this population in relation to national modernization, and it indicates how sertanejos themselves understood the drought problem and federal efforts to correct it. The subsequent chapters focus on the role of technical personnel in debates about sertão development. Each considers a different professional group that was granted institutional authority to make recommendations about or oversee sertão development during the period discussed. As a series of technocratic cohorts gained influence over government planning in the sertão, each portrayed its profession as offering essential modernization expertise. Backed by the authority of international advisers in their fields, these Brazilian scientists worked to adopt an apolitical posture in their policy recommendations. They believed that the specialized knowledge of their particular fields would generate similar outcomes even when applied in contexts subject to very different political dynamics. This was the essential promise of science: its universal applicability. Chapter 2 examines the rural health campaigns led by Brazilian and American sanitarians from 1910 to 1930. Chapter 3 looks at the drought agency's activities, led by civil engineers, for the years 1909–30. Chapters 4 and 5 consider the growing but ultimately modest impact of agronomists on the drought agency's agenda during the period dominated by Vargas's administrations, 1930–55. Chapter 6 turns to the revised regional development strategy proposed in the years 1948–64 by economists who directed new agencies established to supplant the national drought department's authority in the sertão.

· · · · · ·

The impact of the dams, roads, and irrigation canals constructed in the sertão from 1909 through the 1960s was primarily to solidify long-standing social dynamics, intensifying landowners' control over natural resources and the human beings who depended on them and increasing their power as political patrons. Federal agents working in the sertão remained largely beholden to traditional rural power brokers, and landowners' priorities heavily influenced regional development agendas. Technocrats' scientific expertise and training never trumped the material authority and political clout of landowners and industrialists. Rather than providing a peaceful means to circumvent regional tensions while achieving social change, the infrastructural improvements provided by northeastern Brazil's development personnel reinforced existing lines of social fracture. Alternative outcomes would have required more direct confrontation with the status quo, which the drought agency's reigning bureaucrats and their political allies were unable or unwilling to pursue.

The history of drought control efforts in northeastern Brazil suggests a general critique of technologically based development engaged in by many governments and nongovernmental organizations during the twentieth century. Despite the recurrent optimistic belief of technocratic personnel that their skills could solve the entrenched problems of historically poor regions without engendering massive social upheaval, there was often no way to bypass political confrontation and still remedy the long-standing marginalization of impoverished regions and people. Technocrats operate in a political landscape that shapes the potential impact and effectiveness of their recommendations. As anthropologist Tania Li observes, however, their objective is to represent intractable problems as solvable through technology—to "render [them] technical."[9] This means excluding, from their definition of the focal problem, elements such as political economy or social inequality that will not respond to a technical solution. As historians of technology have demonstrated in myriad cases, the power of technology to effect change is constrained by the social context in which it functions. By overlooking crucial aspects of this context, development technocrats impeded their ability to achieve the goals that many sincerely aimed for. This failure made their work more palatable to governing elites than it would otherwise have been, and thus helped to perpetuate their employment as regional developers.

# 1  Climate and Culture
## Constructing Sertanejo Marginality in Modern Brazil

Drought and the human suffering that accompanies it have been central to popular images of Brazil's northeastern hinterland since at least the 1870s. Yet the designation of the Nordeste (Northeast) as a distinct region emerged only after Brazil's federal government established the first drought works agency in 1909 to address the periodic affliction in its semiarid zone. Prior to that time, the usual geographic designation for the states from Bahia to Ceará was simply "the north." The brief economic flourishing of the Amazon basin during its rubber export boom led to the widespread adoption of different terms for the two major subregions of the north: "the Amazon" and "the northeast." Thus, from the time the term northeast was first adopted, it was strongly associated with drought and chronic poverty, as well as with sugarcane cultivation on the humid coast.[1] The term sertão had been used by early Portuguese settlers to describe all unexplored inland territories of their new colony. Yet by the 1910s the term was increasingly reserved to refer specifically to the northeast's drought zone.

Due to the dismal portrayals of hardship and struggle that came periodically to southern Brazilians' attention, Brazil's newly delineated sertão was inextricably linked in the national imagination to prevailing concerns about racial degeneracy and cultural backwardness. Sertanejos, as the region's inhabitants are known, were understood by many elites—especially in the southern and coastal economic centers—to have been deeply tainted by centuries of racial intermingling and a pervasive culture of political corruption, violence, and feudal inequalities. Twentieth-century development of the sertão was fraught in part because of the widespread perception that it was culturally and climatically doomed. This chapter introduces the history of the sertão in the late nineteenth and early twentieth centuries, emphasizing tensions between Brazil's modernizing ambitions and the construction of the region and its inhabitants as chronically backward. Development efforts, particularly during the early decades of the federal drought agency, must be understood in the context of regional power dynamics and national perception of the intersections between the sertão's climate and its culture.

## The Geography and Economy of the Sertão

The sertão's normal climate cycle consists of a rainy season (generally the first half of the calendar year) and a dry season. Some areas typically receive precipitation for only three months of the year. Occasionally part of the sertão experiences what is known as *seca verde* (green drought), in which the usual annual precipitation falls during a very brief period. A drought is generally declared when little to no rain has arrived in an area by mid-March, several months into the usual rainy season. This phenomenon is understood to be triggered by interactions among major wind currents off the Atlantic. Droughts have been recorded since the late sixteenth century, and the sertão's characteristic *caatinga* vegetation is adapted to withstand lengthy dry periods.

The spiny plants of the caatinga have narrow leaves with waxy coatings that limit evaporation of moisture from their surface. If the leaves and grasses do wither, cattle must be moved from the drought zone or they face starvation. As early as the colonial period, large ranchers in the sertão accumulated "reserve" properties outside the semiarid zone where they aimed to move their livestock during droughts.

During the decades covered by this study, widely recognized drought years occurred in 1915, 1919–20, 1931–32, 1942, 1951–53, and 1958. What constitutes a severe drought in terms of pleas for aid, media coverage, and political response depends more on the event's human impact than on any hydrological or geographic measurement. It is not possible to "read" the widely recognized drought years directly from standardized references such as the Palmer Drought Severity Index (developed in the 1960s to record periods of reduced or excessive precipitation across global regions). Instead, times remembered in the northeast as severe droughts are those in which reduced rainfall was experienced for a long time—thus causing significant food scarcity—or across a heavily populated area, resulting in hardship for large numbers of people and livestock.

This harsh environment has been on the literal and metaphorical margins of Brazil since the Portuguese founded their colony. In the early sixteenth century, when sugar plantations began to be established in the northeast (around the colonial capital of Salvador da Bahia), bands of indigenous people were forcibly "descended" from the hinterland sertão by trad-

Sertão landscape, taken during the expedition led by Belisário Penna and Arthur Neiva in 1912. Source: Imagem BP (F-VPP) 1-22, Acervo da Casa de Oswaldo Cruz, Departamento de Arquivo e Documentação.

ers to work as slaves.[2] As fields of sugarcane expanded along the cleared forest land of the northeast's humid coast, ranchers began to push into the sertão to pasture their cattle herds. These beasts were essential to the sugarcane economy, providing muscle power for mills, milk and meat protein to sustain the workforce (which by the late sixteenth century consisted primarily of African slaves, as indigenous workers had escaped or died from disease and exhaustion), manure used as fuel, and leather for nearly every purpose imaginable. During the seventeenth century, cattlemen followed the São Francisco River inland, gradually driving their herds north of its banks and along smaller inland waterways. The crown readily granted large tracts of land to loyal subjects willing to colonize the frontier by laying claim to former indigenous territory (and exposing native Brazilians to lethal pathogens that hastened their decline). Each such grant (called a *sesmaria*) comprised several thousand hectares of land, and an individual could acquire many over his lifetime. These properties were unfenced; grazing boundaries were enforced through custom and, at times, violence.

Until the mid-twentieth century, ranch hands were often paid partly in livestock, and this allowed them to amass a herd over time, a potential route to independent ranching. Historians have estimated the minimum sustainable ranch in the sertão to be one thousand hectares, sufficient for 50–150 head of cattle.[3]

When gold deposits were discovered in Minas Gerais in the 1690s, the northeast's ranches began providing meat and muscle to that expanding industry as well as to the coastal sugar plantations of Alagoas, Bahia, and Pernambuco. The rise of salted beef production in the eighteenth century helped cattlemen in Ceará and other provinces distant from the flourishing mines to compete with southern ranchers close to Brazil's expanding port city of Rio de Janeiro (which had been designated the new colonial capital in 1763—the result of its central role in gold exports). But a severe drought in the 1790s that decimated *cearense* cattle provided an opening for ranchers in the far southern province of Rio Grande do Sul to expand their own provision of beef and leather goods to Rio de Janeiro, where coffee production and exportation were on the rise.[4] These southern cattlemen benefited from the fact that their lands were not subject to the droughts that periodically devastated livestock in the sertão.

As this history indicates, notwithstanding the fact that the fertility of sertão pastures (and the consequently easy life of its ranchers) have been romanticized by northeastern boosters since the nineteenth century, the region is not actually well suited to raising cattle. It was simply used for that purpose during the heyday of Brazil's sugar exportation, when sertanejo cattlemen benefited from their proximity to the colony's most thriving region. As geographer Kempton Webb has noted, "The sertão is not really very good for cattle *or* sugarcane, but cattle can survive there and sugarcane cannot."[5] The sertão thus provided a welcome economic opportunity for men on the margins of the colonial sugar economy who were willing to risk migrating into the dry hinterland in hopes of increasing their fortunes and living independently of coastal plantation owners. When German naturalists Johann von Spix and Karl Friedrich Philipp von Martius explored parts of the northeast in 1800, they estimated the sertão's population at 726,000.[6] By 1915, cattle herds topped 1.5 million, and the sheep population was double that—all grazing on open pasture.[7]

Agriculture has been the other backbone of the sertão's economy, though rocky soils and the threat of drought limit cultivation over much of the region. Extensive ranching may have diminished the viability of agriculture in parts of the sertão because cattle were commonly moved to river margins and more fertile areas during dry periods, damaging these scattered patches of nutrient-rich soil. Nevertheless, farming expanded during the late eighteenth century in response to Britain's growing cotton demand and the need for food crops within Brazil itself. Ranchers may have taken up farming during this period because of the growing competition from cattlemen in Rio Grande do Sul. The expansion of U.S. cotton production after 1820 led some sertanejo farmers to diversify into such products as carnauba palm wax, which was better able to compete on the global market. In the 1840s, the impact of drought on cattle herds propelled additional cearense ranchers to begin farming, cultivating for both export and local subsistence.[8]

The growth of the sertão's agricultural economy, which drew migrants from the northeast coast to the interior, meant that new settlers farmed increasingly marginal areas. Many of these migrants were squatters. The Land Law of 1850, which required that new land titles be issued via sale only (in keeping with nineteenth-century economic liberalism), was an attempt to eliminate this practice throughout the country, but it was largely ineffective—both in the sertão and elsewhere. Although there is little detailed scholarship on land settlement in the sertão as a whole, Martha Santos has conducted a careful and revealing study of landholding in several backland municipalities of nineteenth-century Ceará, the province with the largest portion of territory in the sertão. Ceará's small farmers enjoyed a period of economic security and autonomy from 1840 to 1877 due to increased demand for agricultural goods and the disruption in U.S. cotton exports caused by the U.S. Civil War. As a result, Ceará's population more than doubled, growing from 350,000 in 1850 to 817,000 by 1877. Most newcomers to the area were *pardos* of mixed African and European ancestry from the northeast coast. Nineteenth-century census data for Ceará confirms that just over half of the state's population claimed some African ancestry as *pardos*, free blacks, or slaves.[9]

The devastating "Great Drought" of 1877–79 contributed to a significant downturn in the fortunes of Ceará's sertanejo farmers. Once families depleted modest food reserves, they undertook arduous journeys toward the coast, traveling many miles on foot in search of food and assistance. Outside Fortaleza and other capital cities of the Nordeste, drought refugees were held in unhealthy and overcrowded "concentration camps" (as they were termed by the government) until they could return home; in these squalid encampments many succumbed to infectious disease. Thousands were shipped at government expense to unaffected provinces or to Amazonian rubber fields, in search of charity and whatever paid labor they could find.

Santos calls the 1870s "calamitous years" for Ceará's small farmers because of the multiple changes that combined to gravely reduce their household security. Food prices had risen since the 1850s, as the northeast's farmers turned increasingly to more profitable export crop production. Military impressment during the Paraguayan War (1864–70) meant that many families lost valuable laborers to either military service or flight from it. After a brief boom in Brazilian cotton exports during the 1860s, U.S. plantations returned to cotton production, displacing the arboreal Brazilian crop less favored by British buyers. This decline in cotton demand left many northeastern farmers with significant debts, which they often repaid by selling slaves to prosperous coffee farmers in São Paulo (who could no longer obtain slaves through the Atlantic trade after 1850, due to British abolitionist pressure and naval muscle). Thus when the Great Drought struck toward the end of the decade, sertão small farmers' wealth in land and labor had already begun to decline.

Santos also finds that the size of many families' estates in the sertão declined from the colonial period to the mid-nineteenth century. This was due to Brazil's inheritance laws, under which parents' property and other assets had to be divided equally among all legitimate children, male and female. Although strategic marriages among wealthy families worked to reconstitute vast estates, the overall trend was toward fragmentation. By the 1850s, less than 10 percent of properties in the cearense municipalities that Santos analyzes were on the scale of colonial era *sesmarias*.[10] The Great Drought, however, reversed this pattern and led to a reconcentration of land and wealth in the parched sertão. As the 1877 harvest failure compounded the cotton slump, panicked merchants demanded immediate payment or forfeiture of collateral property by sertanejo farmers, whom they feared would never manage to repay them. As property holdings diminished due to partible inheritance or repossession by creditors, farmers were decreasingly able to sustain their households and were often forced to sell their remaining land. Property values plummeted by as much as 80 percent during the extended drought, offering a tantalizing investment opportunity for the few landowners able to shoulder such risk.[11]

Santos thus finds a rise in the proportion of cearense landowners (at least in the municipalities she studies) who claimed two or more property holdings, from 57 percent in 1860 to 73 percent twenty years later after the Great Drought. This represents a transfer of wealth from those most afflicted by the natural disaster and economic downturn to those better positioned to withstand both. Her research indicates a rise after 1880 in dependence on landholding patrons among "free poor farmers and ranchers [in Ceará] . . . no longer able to benefit from the smallholding pattern [formerly] characteristic" of the sertão.[12] These once autonomous farmers became day laborers, earning subsistence wages with little possibility of acquiring assets to pass along to their children. The Great Drought, coming at a time of economic instability across the sertão, thus contributed to multigenerational impoverishment. It made the least secure sertanejo families even more vulnerable to famine and forced exodus during future drought years.

## Sertanejo Marginality and National Modernization

Many compassionate portrayals of the drought exodus in novels, popular songs, and folk poetry emphasize sertanejos' reluctance to leave their native land despite the periodic hardship they experience there. Most families in such accounts abandon their farms only after exhausting every possible food source, at which point they rely on knowledge of edible plants for nourishment along the precarious route to the more humid coast. Sudha Swarnakar observes that the customary choice of the word *retirante* (drought migrant) to describe drought refugees emphasizes the sertanejos' perpetual longing to return home and their lack of a clear or permanent destination. She notes the difference in meaning between a *migrante*, "one who chooses to migrate as a better option for his future," and the drought *retirante*, "one who gives up or withdraws" as a temporary escape from misfortune.[13]

Sertanejos' deep attachment to their native region puzzled the many politicians and drought agency engineers who proposed mass resettlement of the northeast's backlanders to more hospitable areas of Brazil during the twentieth century. This incomprehension is reflected in a poem, written at the turn of the twentieth century, about drought victims fleeing the state of Ceará. Poet Cordeiro Manso, from the neighboring northeast state of Alagoas, paints a dismal picture of the mass starvation against which sertanejos struggled. Alagoas had been ordered by the federal government to accept a shipment of one thousand drought retirantes, and Manso advises his fellow statesmen to receive the migrants graciously. But his verse expresses incredulity that cearense sertanejos would willingly suffer so much before leaving their homes. In Manso's narrative, a family of retirantes abandons its emaciated cattle to marauding jaguars and flees toward the coast, carrying their smallest children and a few possessions. Two children die of hunger along the way, and the protagonists see a family trade a child to work as a slave in return for a few cakes of raw sugar (*rapadura*). The drought victims are aware that the northeast's governors and bishops have been given aid from the federal government, but they know that this will trickle through the ranks of the powerful without reaching the starving poor. Manso concludes by asking,

| | |
|---|---|
| *Quisera saber ao certo* | I'd like to know for certain |
| *Se quando um dia chover* | If one day when it rains |
| *Se o povo cearense* | The people of Ceará |
| *Volta ao lar que o viu nascer,* | Will return to the place of their birth, |
| *Dando glórias ao sertão* | Praising the sertão |
| *Que lhe fez tanto sofrer!* | That made them suffer so much! |
| *No meu modo de pensar* | By my way of thinking, |
| *Dando a minha opinião,* | To offer my opinion, |
| *Eu não dou meia pataca* | I would not give two bits |
| *Por dez léguas de sertão* | For ten leagues of the sertão |
| *Eu corro com medo dele* | And would run from fear of it |
| *Como quem corre do cão.* | Like someone running from the devil.[14] |

To understand the conundrum of sertanejos' fierce devotion to their native land, despite its climatic and social shortcomings, one must comprehend how foreign they often feel in other parts of Brazil. For much of the nineteenth century, the risks of drought years were offset by the independence that the sertão offered small farmers and ranch hands at all other times. Sertanejos who migrated to coastal cities in the northeast and elsewhere were relegated to the social margins. This was due both to their mixed racial heritage and to significant disparities in wealth and influence between the northern and southern regions of Brazil by the 1870s. Many southern Brazilians, particularly those who claimed substantial European ancestry, viewed poorly educated, often illiterate sertanejo migrants as hopelessly backward.

Most sertanejos are descendants from native Brazilians (*índios*) who remained at the frontiers of white colonization; European settlers, primarily the Portuguese and Dutch (due to a mid-seventeenth century Dutch occupation of Bahia and Pernambuco); and black or *mestiço* refugees from coastal plantation society. Brazilian elites of the late nineteenth century were pessimistic about how their country's largely mixed-race population could spawn a modern nation given prevailing European theories (used to support imperial ambitions) that designated whites as the most highly evolved racial group. Brazil's late imperial and early republican governments subsidized European immigration in the 1880s and 1890s to increase the proportion of whites in their population. Many elites believed that blending white Europeans with mixed-race Brazilians through intermarriage, a process termed *branqueamento,* or "whitening," would elevate the country to a higher level of civilization. Because whiteness was such an important marker of modernity during Brazil's early republic, national leaders were often dismissive of sertanejos' potential contribution to their new nation.

Many southern elites supported the formation of Brazil's federalist republic in 1889 because of the substantial economic divide that separated provinces in the north and south of the country by that time. *Paulistas,* in particular, felt that their economically dynamic state (São Paulo) was hampered by the resource demands of the decadent north. They had invested income from booming coffee exports into rail lines and urban infrastructure. The republican constitution of 1891 rewarded São Paulo's coffee planters with greater state autonomy than the centralized imperial system had allowed, while also reducing the northern states' representation in the national legislature. During the republic's early decades, the balance of power in Rio de Janeiro and São Paulo states shifted toward bankers and businessmen who profited from agricultural exports as well as urban industrialists who turned these profits to other productive uses. The south's economy was evolving more rapidly than the northeast's, and periodic droughts reduced the northeastern hinterland's economic productivity even further.

The sense of a widening divide between northern and southern Brazil was felt by citizens in both regions during the early republic. Brazilian historian Durval Muniz de Albuquerque Jr. has observed that a specifically northeastern identity emerged during the early twentieth century as an assertion of regional independence from São Paulo's self-conscious and chauvinistic modernization. Northeastern elites were wary of national policies that favored southern coffee interests and industrialization, threatening to maintain their states in a subordinate position. Such protective regionalism inspired an outpouring of artistic production that celebrated local traditions and folk customs and described particularly Nordestino experiences— including the struggle with drought. By 1930 northeastern regionalism as expressed in folk music and other forms of popular culture was the deliberate antithesis of *paulista* modernism.

## National Perceptions of the Sertão:
## From the Great Drought to Canudos

National awareness of the growing disparity between Brazil's economically dynamic south and the northeast became acute during two late
nineteenth-century crises: the Great Drought of 1877–79 and a prolonged
standoff between a sertanejo settlement and the republic's army in 1896–
97. In both cases, new technologies helped communicate the hardships suffered by sertanejos to southern Brazilians more directly than had previously
been possible.

Estimates suggest that the 1870s drought caused 220,000 deaths across
the northeast through starvation and disease.[15] Ceará was the most impacted province; it lost more than 100,000 inhabitants to death or outmigration over the course of three years.[16] Entire households fled the
interior on foot, packing their few possessions onto a mule. Waves of epidemic disease swept the migrant routes—particularly smallpox, which had
not been seen in the sertão since 1825. The economic downturn and rising
malnutrition since the 1860s had created fertile ground for cholera (which
was sweeping across international trade routes at this time) and other infectious illnesses, and the drought exodus exacerbated these conditions.
When migrants arrived in cities outside the drought zone, they amassed in
makeshift camps where unhealthy conditions alarmed urban authorities.
Ceará's capital, Fortaleza, had a predrought population of 27,000, but more
than 100,000 migrants sought refuge there when the crisis was most acute.
Desperate accounts from the besieged city—possibly exaggerated by elites
hoping to spur an outpouring of government and charitable aid—reported
a litany of vices in and around the encampments, including murder, prostitution, and cannibalism. Attacks by bands of refugees on local stores and
warehouses to obtain food and other essential supplies were a constant
threat to urban property owners. Disease raged in the camps and vermin
multiplied; on one day, 1,004 smallpox victims required burial in mass
graves on the outskirts of Fortaleza.[17] For all of these reasons, public
authorities in the northeast implored the imperial and provincial governments to provide aid. To awaken southern Brazilians' sympathy for
drought victims, photos of the starving were reproduced (often as line
drawings) in popular magazines in one of the first uses of photojournalism
in Brazil. These images and the accompanying stories of wrenching deaths
conveyed the horror of famine and epidemic disease to readers far from the
sertão.[18]

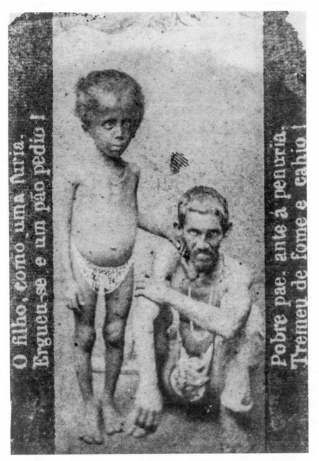

Starving sertanejos depicted in a Rio de Janeiro magazine, 1878. Photograph by J. A. Corrêa. Source: *O Besouro* 20 (1878): 1. Acervo Fundação Biblioteca Nacional.

Nineteenth-century liberal ideology dictated that aid to the indigent be provided in return for their labor whenever possible. This was true in institutions like Recife's Santa Casa de Misericórdia (a hospital, asylum, and orphanage), and it remained true during droughts.[19] Although much aid to drought victims was, of necessity, directed to the cities where refugees gathered (which in turn drew an increasing volume of migrants from the interior, where little aid could be obtained), the goal of urban authorities was to establish public works projects outside of their cities and to redirect drought migrants to those centers of assistance. Road construction was a priority, to speed the transport of water and other supplies into the interior, though even with improved roadways armed gangs remained a threat, as they waylaid mule trains and distributed precious cargo to their own followers. Some sugar mill owners on the coast were persuaded to accept drought refugees as sharecroppers, paid according to a government-stipulated contract (which the mill owners thought overly generous, even as they accepted hundreds of workers).[20] As drought aid became more institutionalized in the twentieth century, it was common for influential men to enlist their political clients in coveted public works projects. The distribution of aid during droughts became a powerful enactment of patronage for the northeast's public officials, and government and church authorities eagerly emphasized their role in obtaining federal largesse. As Santos observes in her study of sertanejo masculinity, "The need to rely on the handouts that others were willing to give as charitable acts contrast[ed] sharply with the celebrated resourcefulness and autonomy that once had defined the honor of successful small farmers and ranchers" in the nineteenth-century sertão.[21]

The 1870s drought caused a greater loss of human and animal life than any subsequent crisis in the northeast, but similar dramas were repeated in 1888 and at the turn of the century. For the republic's reformers, the periodic scenes of drought victims' mass exodus were painful reminders of how far Brazil's sertão lagged behind their modernizing vision. Elites in the industrializing cities of Rio de Janeiro and São Paulo were committed to positivist ideals of rational progress guided by scientifically trained professionals. The northeast, with its domineering landowners, minimal state presence, flagging agricultural economy, and desperate migrants, appeared increasingly backward when compared to the country's political and economic core.

The second episode that brought the sertão and its inhabitants to national attention during the late nineteenth century occurred at a squatter settlement in the interior of Bahia. Charismatic religious leader Antônio Conselheiro established the town of Canudos near two seasonal rivers (only the São Francisco River runs year-round in the northeast).[22] The "counselor" began on good terms with neighboring landowners, but his opposition to the republic's tax laws soon irritated government authorities. Conselheiro advocated a return to the monarchy as a more protective form of government; this appealed to many sertanejos who perceived laws enacted during the republic's early years as abnegations of the state's traditional paternalistic role. Vagrancy laws were enforced to enlist "idle" workers as replacements for the diminishing rural labor supply following abolition (enacted with little fanfare in 1888). Combined with a despised 1874 law allowing military impressment of backlanders, this intensified government infringement on sertanejos' liberty. Additionally, in response to the greater administrative autonomy devolved to them by the republic's federalist constitution, states levied new land taxes to raise revenue. This led to increases in the rents owed by tenant farmers, an added hardship for an already impoverished population.[23] Sertanejos were also deeply unsettled by the secularizing efforts of liberal republican positivists—particularly the institution of civil marriage, which aimed to wrest authority over individuals' private lives away from the church. Meanwhile, landowners battled violently in the backlands for control over rural municipalities, which had become politically significant under the republic's decentralized administrative system. Many sertanejos feared this unstable new political and economic order, with its apparent indifference to their material and spiritual well-being.

By 1895, when the Canudos settlement's several thousand households constituted Bahia's second largest city, federal and state administrators perceived it as a threat to their authority. Following a skirmish between Conselheiro's disciples and local police, federal soldiers attacked the compound. The ragtag *canudenses* staved off this and several subsequent military assaults. War correspondents for southern newspapers transmitted eyewitness reports of the conflict via telegraph, the first time this had been possible in Brazil's military history. Their accounts allowed readers throughout Brazil to follow the events at Canudos in dramatic and gory detail as they unfolded over the course of two years. The federal offensive at Canudos ended after four campaigns with a brutal massacre of every man in the settlement and many women and children as well. As historian Dain Borges notes, rural elites in Bahia were already keenly aware that paternalism wove a fragile bond between rich and poor. The Canudos rebels confirmed landowners' worst nightmare: the rise of a defiant and ungrateful underclass organized to oppose elite authority. Such fears fed the desire to crush Conselheiro's dissident band as a lesson to other disloyal clients.[24]

In 1902, a thirty-six-year-old engineer-turned-military correspondent published a vivid recounting of the federal army's Canudos campaign that became a classic of Brazilian literature.[25] Euclides da Cunha spent part of his childhood in Bahia, in 1884 went to Rio de Janeiro's Escola Politécnica, and then attended the Escola Militar where he trained as a civil engineer. There he was taught by Brazil's archpositivist, Benjamin Constant, during the period of republican foment, and he absorbed Constant's devotion to scientifically based social reform. During a brief dismissal from the army for insubordination in 1888 da Cunha worked as a journalist, and he left the military completely in 1896. By then he had grown disillusioned with the authoritarianism of the republic's early leaders. Da Cunha covered the

Canudos rebellion for the *Estado de São Paulo* newspaper, and his interpretation of that conflict reflected his own concerns about Brazil's lack of national unity and the republic's neglect of education, technological infrastructure, and administrative oversight in its vast hinterland.[26]

Da Cunha's epic narrative of the Canudos uprising, *Os Sertões* (translated into English as *Rebellion in the Backlands*), was compiled from his contributions to the *paulista* paper during the war and embedded in a larger framework emphasizing its national significance. After witnessing the tragic end of the Canudos rebellion, da Cunha was deeply troubled by Brazil's inexorable and self-destructive march toward modernization. He saw both sides in the battle for Canudos as barbaric—the sertanejo rebels for their inferior racial character and lack of education, the military for degenerating into cruel vengeance as it struggled to defeat Antonio Conselheiro's fiercely loyal band. Da Cunha described his country as "condemned to civilization" through a wrenching process that would eventually fuse a single Brazilian race from several disparate original stocks. *Os Sertões* brought home to many readers the challenges that Brazil's new republic faced in drawing the backlands into its modernizing project, and the book was immediately hailed by elites as a masterpiece. In the year following its publication, da Cunha was admitted to the Brazilian Academy of Letters and the Institute of History and Geography, the country's two preeminent scholarly institutions.

In da Cunha's depiction, the sertanejo's well-honed ability to read his environment aided survival both during droughts and when under attack by federal troops. The Canudos conflict began in November 1898, during the dry season. While federal soldiers had difficulty marching in the scorching heat or navigating the arid caatinga scrub, their better-acclimated foes were adept at sustaining themselves and their horses, even in apparently barren terrain. *Os Sertões* often describes sertanejos' integration into the surrounding landscape as if they were simply another xerophilous species adapted to withstand the semiarid climate. Like the region's native plants, sertanejos conserved resources until survival depended on them, exhibiting remarkable energy and tenacity when circumstances demanded. Later writers frequently adopted this naturalized depiction of sertanejos, romantically attributing such ecological synergies to their indigenous heritage.

In discussing drought as a pervasive feature of sertanejo life, da Cunha describes two rituals commonly engaged in to predict rainfall. His portrayal of these practices plays upon stereotypical characterizations of the sertão as a land of fatalistic superstition and mysticism. On December 12, the eve of St. Lucia's Day, sertanejos set out six salt cubes representing each of the six months of the usual rainy season, January to June. The next morning, any cube that has dissolved forecasts rain during that month. If none have dissolved or absorbed significant condensation, a drought is predicted. Da Cunha suggests that this exercise may have some scientific merit, since it measures the level of atmospheric vapor a few weeks before the rainy season should commence.[27] The other important date for drought prediction among sertanejos is March 19, the Feast of São José. This is widely believed to be the last possible day for rains to begin; if they have not started by then, many sertanejo farmers begin bracing for a drought year.

Da Cunha's descriptions of sertanejos in *Os Sertões* played upon existing stereotypes of the Euro-Indian *caboclo*, a national type imbued with both positive and negative associations. Romantic northeastern authors of the late nineteenth century, like José de Alencar, celebrated the racially mixed *caboclo* for his distinctive ethnic character and unique psychology. Da Cunha also depicted racial intermixing as having been beneficial, rather than degenerative, in the backlands—producing a racial type adapted to the harsh environment. He warned readers, however, that the sertanejo's peculiar traits, the product of long isolation and inbreeding, posed an obstacle to governing the primitive sertão from the more cosmopolitan coast. Sertanejos had acclimated to the demands and constraints of their merciless landscape and could persist there admirably, he believed, unless threatened by a civilization that had evolved far beyond them, as happened at Canudos. These views were consistent with social Darwinian ideas about racial difference, evolution by natural selection, and degeneration prevalent among Western intellectuals in this period.

Despite his emphasis on the sertanejos' alien character, da Cunha sought to rescue them from the harsh assessment that European racial theories leveled. In his view the Canudos rebels were a truly Brazilian race, adapted to the country's natural conditions and not indebted to another culture for their values—unlike coastal society, which emulated Europe. Like many other Brazilian intellectuals in this period, da Cunha adopted a Lamarckian view of race and evolution in which social processes and life experiences could influence later generations by altering heredity.[28] He believed that the sertanejos could be united with coastal society through a deliberate civilizing process, using education and civil engineering to bring them the material and technological advantages that the more advanced areas of Brazil already enjoyed. In da Cunha's view the challenge of the sertão centered on its physical environment more than on its people; and the careful introduction and management of improved technology would alter both the landscape and its inhabitants. Thus his work was ultimately redemptive, intended as a lesson to republican leaders who had horribly botched their recent encounter with recalcitrant backlanders.

Historians have offered several corrections to da Cunha's analysis of the events at Canudos. Robert Levine notes that the portrayal of canudenses as generic *mestiços* with no predominant racial heritage was inaccurate; a number were discernibly Afro-Brazilian, and some were probably recent immigrants from Italy or Portugal. By overlooking this immigrant population, da Cunha and other reporters exaggerated the cultural and biological gulf that separated coastal Brazilians from inhabitants of the interior—a central element in their framing of the events.[29] At the same time, he may have deliberately downplayed the African ancestry of many sertanejos in order to elicit greater empathy for them from elite Brazilians accustomed to embracing the mixture of indigenous and European elements in their national body while overlooking or denigrating its substantial African heritage.

Many republican interpreters of the Canudos war depicted Conselheiro and his followers as atavistic religious fanatics, victims of messianic delusion.[30] Contemporary historians reject such portrayals; instead they interpret the encampment at Canudos as a pragmatic response to changes wrought by the new governing regime. Levine critically analyzes the *visão do litoral* (coastal vision) that led metropolitan observers of the Canudos episode to misrepresent sertanejos' motivations. He asserts that Conselheiro's followers were neither lunatic nor utopian, neither messianic nor revolutionary, and argues that Conselheiro established a stable community based on disciplined religious devotion, which appealed to its inhabitants during an uncertain and frightening time. Tempering purely material interpretations of Conselheiro's appeal, Patricia Pessar emphasizes the millenarian worldview that led many Nordestinos to see the political, economic and social disturbances during this period in apocalyptic terms.[31]

Republican analysts of the canudenses' resistance, however, were not conditioned to interpret the sertanejos' resolve in light of the material and symbolic significance they assigned to recent changes in government. As Stanley Blake notes, "Psychological explanations of individual and collective human behavior were far more palatable than economic or political explanations which would have brought attention to the social and economic inequalities of Brazilian society. To argue that the rebels were justified or that their demands were rational would have been too controversial for regional and national elites."[32] Historian Gerald Greenfield highlights the ignorance and romanticization of the northeast which shaped elite response to Canudos. It was widely believed by coastal intellectuals (most of whom had never visited the sertão) that the land's aridity had been caused by primitive agricultural practices involving burning to clear fields.[33] Da Cunha himself adhered to this theory, describing the sertão's barrenness as the result of Indian and Portuguese settlers' routine use of fire to create pasture, combined with a ferocious wind and rain cycle.[34] Such criticism implied that more rational cultivation and pastoral practices in the sertão would end the region's climatic woes and improve the backland economy. Baseless speculation about the sertão's natural fertility led to accusations of sertanejos' laziness as the cause of their crippling poverty, an idea easily lifted from prevailing European racial theories about "tropical torpor." Greenfield argues that such late nineteenth-century perceptions conditioned the national response to Conselheiro and his followers: "Given prevailing social class attitudes, the specific images of the northeast backlands, and an already established vocabulary for describing popular movements as threatening and anarchic, could the people who flocked to Canudos be anything other than rude, ignorant, superstitious masses caught in the grip of a charismatic fanatic? Sertanejos acting in this fashion could only be perceived as the antithesis of civilized society—the order and progress promised by republican Brazil."[35] Thus northeastern backlanders, already viewed by many elites as racially suspect, came increasingly to be seen as obstacles to rational, progressive modernization.

Nevertheless, in response to da Cunha's account of the events at Canudos, some early twentieth-century Brazilian nationalists expressed more moderated views of the sertanejos' place in the modernizing nation. A number of republican authors portrayed them as noble and courageous people unfairly neglected by their compatriots; several wrote with great empathy about sertanejos' suffering—particularly during the drought migration. A lecture read to Rio de Janeiro's prestigious Instituto Histórico-Geográfico Brasileiro during the 1919 drought, for example, emphasized sertanejos' tenacity in confronting repeated hardships and their resourcefulness in surviving harsh months by burning the spines from cacti so they could feed the cacti to their livestock.[36]

Speaking at a commemoration of da Cunha's work ten years after his untimely death (he was killed in 1909 by his wife's lover, a fellow soldier), anthropologist Edgard Roquette-Pinto dismissed da Cunha's fear that Brazil's formative backlands race would be driven extinct by modernization. Roquette-Pinto asserted that da Cunha's hierarchic understanding of racial difference reflected his exposure to the writing of bigoted foreign scientists like Harvard University professor Louis Agassiz.[37] In the anthropologist's more optimistic view, nations need people with different traits suited to a variety of environments and circumstances. Brazil's *jagunço* (a generally pejorative term for sertanejos that was used frequently in reports of the Canudos rebellion) was as essential to the country's future as the south's gaucho and immigrant Japanese farmer—provided that these groups could all be united through a shared national culture. Roquette-Pinto's emphasis on culture, environment, and education as more influential than biological race reflected a change in anthropological thought in Europe and North America during these decades. His argument for the richness and utility of Brazil's regional differences became central to nationalist discourse of the late republican period and the subsequent administrations of Getúlio Vargas, particularly among Nordestino regionalists like Gilberto Freyre. Prominent intellectuals of the early republic, including Manoel Bomfim and Alberto Torres, also described national unity as a sociological rather than racial achievement, diminishing the "whitening" agenda so dear to many of their southern compatriots. Instead of advancing by attracting European immigrants, Torres and others argued, Brazil's leaders should use education and modern technology to civilize Brazil's rural populations, which had been unjustly abandoned to debilitating environments. In this interpretation, environmental remediation guided by modern science would redeem a backward landscape and its long-suffering population in the interest of national progress.

## Drought Alleviation and Political Patronage

Scattered attempts to mitigate drought in the sertão were made during the nineteenth century, particularly following the Great Drought. Proposals included reforestation to increase rainfall (what historian Roger Cunniff terms the "rainmaker" school[38]) and connecting the northeast's most significant perennial waterway, the São Francisco River, to Ceará's Jaguaribe riverbed by canal. Neither of these projects was undertaken, although reforestation received renewed attention in the 1930s when Vargas established an agricultural service within the federal drought works agency.

The projects that did receive funding involved reservoir construction. Among the strongest advocates for this approach was influential cearense senator Thomas Pompeu de Sousa Brasil (active in Ceará's Liberal Party from the 1840s until his death in 1877), whose own family had been forced

Quixadá Reservoir in Ceará, photographed during a 1912 expedition. Source: Imagem IOC (AC-E) 7-18, Acervo da Casa de Oswaldo Cruz, Departamento de Arquivo e Documentação.

to abandon its substantial estate during the drought of 1825 and move to a neighboring town.[39] In the early 1880s the imperial government contracted British engineer Julian Revy to create three large inland lakes, but Revy's work was opposed on several fronts: nationalist resentment of imported engineering talent; criticism from northeastern politicians outside the sertão who favored extending rail lines throughout the region instead; and growing accusations in the national legislature that funds sent for drought aid were unwisely directed to projects benefitting only the landowning elite. Within a few years, Revy's projects were discontinued. Construction of one of his dams, Quixadá, resumed in subsequent decades and was finally completed in 1906. The thirty-year gestation of Quixadá dam confirmed southern Brazilians' suspicion that drought aid was used inefficiently and that many northeastern officials responsible for its disbursement were corrupt.

Historians have confirmed that infrastructural improvements supported by drought aid during the late nineteenth century were often selected to suit the interests of those in power rather than of the population as a whole. For example, several railroad extensions were funded in Ceará during the

1880s, theoretically to transport relief supplies from the coasts to the interior when droughts threatened. Yet, as Greenfield notes, "The relatively minimal positive impact of these lines during the next large drought a decade later confirms the limited degree to which considerations of present and future drought relief figured in the development of the region's railroad network."[40] There is substantial evidence that relieving the plight of sertão migrants was not the primary motive of politicians who campaigned for relief during the Great Drought. In legislative debates, state representatives often described migrants from the interior as a threat to public security. Considerable sums were spent on jails and on guarding food supplies that were vulnerable to attack by starving mobs rather than on improving the conditions of the destitute. And drought migrants served as cheap labor for government-funded dam and road construction often located on private property. Citing Lord Salisbury's use of famine victims for road and port construction in British colonial India as a precedent, northeastern elites enrolled half-starved sertanejo men to work in return for minimal food rations. Since payment for these projects was generally late in arriving from Rio de Janeiro, coastal traders provided food and other supplies without immediate reimbursement. They charged interest for this service, which was deducted from aid money when it did arrive.

Success in obtaining development funds following the Great Drought alerted northeastern elites to the potential for mobilizing national resources in response to the periodic crisis as a way of righting infrastructural imbalances between their region and southern states. Minority political contingents in the region called foul as they watched relief commissions become machines for granting favors to loyal clients (though other Nordestinos countered that this was characteristic of all Brazilian politics).[41] The assertion that only a fraction of aid money reached drought victims became standard rhetoric among politicians from other regions who resented the northeast's drain on the national treasury, to which states like São Paulo contributed disproportionately. Albuquerque has termed drought "an invention through which the [Nordestino] elite aimed to reconquer their position in the national structure and bring about the conditions necessary to perpetuate their longstanding exploitation and domination of that part of the country."[42] While this seems an overstatement given the real misery that droughts inflicted, framing pleas for federal assistance as aid to the neediest sertanejos was unquestionably a self-interested strategy on the part of elites.

The inequities of Nordestino society were perpetuated by a system of political patronage known as *coronelismo*, which has long been a feature of

Brazilian political life. It is most tenacious in the rural northeast, where landowners historically exercised tremendous power over local affairs. The term *coronel* (colonel) refers to the rank held by landowners in the imperial National Guard established in the nineteenth century (primarily to quell slave rebellions). Over time this title came to be used in rural areas as a general term of respect for men of influence. In competition for limited public funds, coronéis acted as intermediaries between their dependent "clients" and the state, in its multiple forms.

With the establishment of the federalist republic in 1889, the central operating mechanism of coronelismo became what political scientist Victor Nunes Leal has termed a "politics of the governors." The 1891 constitution made the state Brazil's predominant political unit; most states usurped authority over municipal fiscal administration within a few years on the grounds that municipalities—particularly, rural ones—were incompetently managed. Governors forged a reciprocal relationship between their political party and municipal leaders, in which each helped the other to win votes. State administrators granted favors in the form of money or public offices to rural patrons who successfully delivered votes to the party in power. These favors were dispensed to the coronel's clients, maintaining his prestige and gaining votes for him in turn. A coronel who failed to support the sitting governor's party would find himself unseated by a rival whom the state backed with promises of monetary assistance and access to bureaucratic power.

Coronéis retained their social position primarily by marshaling votes, which required them to expend considerable effort on behalf of their clients. Leal and later analysts have described in detail the ways in which twentieth-century coronéis won the loyalty of their supporters. Raymundo Faoro emphasizes that a coronel often adopted the role of *padrinho*, a paternalistic form of friendship in which he looked after the welfare of clients' families—and was the only influential person doing so.[43] A fundamental quality of coronéis was charisma, the ability to inspire loyalty. Faoro and others note that coronéis were not necessarily estate owners, though they were always people of notable wealth or education; they included merchants, bureaucrats, and professionals with influence over public institutions, acting as intermediaries for the poor.

Eligible voters traditionally pledged their electoral support as a matter of honor, a form of repayment for the coronel's mediation on their behalf with bureaucrats and other authority figures such as lawyers, doctors, and employers. As Richard Graham argues in his description of patronage in nineteenth-century Brazil (prior to the introduction of secret ballots), the process of winning votes legitimated social hierarchies, releasing the frustrations of the underclass through the selective granting of favors in return for their electoral loyalty. In the absence of widespread citizen education, few rural Brazilians in the nineteenth or twentieth centuries questioned their dependence on a patron for favors. Rural bosses enabled voting itself by helping clients to register and obtain necessary documents such as birth certificates, providing time off from work obligations, and transporting voters to polls. When negotiation for legal votes failed, state and federal authorities generally turned a blind eye to corruption if it served their electoral interests. Coronéis were granted tacit authority over local police, who employed a variety of tactics to intimidate opposition candidates and their supporters. This threat of violence as a strategy of last resort underlies the coronel system. Besides discouraging protest candidates, it is an incentive for local bosses to support the state, which could otherwise turn the police (technically, state employees) against the coronel and his clients.

During the twentieth century, various federal laws aimed to secure citizens' voting rights by stipulating who could register to vote, improving ballot secrecy, and regulating vote counting.[44] But the coronel system, for all its undemocratic features, relies primarily on legal votes. Meaningful reforms to release clients from their dependent status would need to prevent state governors from disbursing payments only to municipalities that support them, and to assist voters in exercising their rights without help from a local patron. Yet the federal government has itself benefited from coronelismo and participated in it. At points of regime change, such as in the 1930s, when Vargas assigned federal interventors to replace state governors, or in 1964 following the military coup, coronéis were relied upon to win rural voter support for the new power structure. At times the military supplied arms to insurgents intent upon overthrowing governors whom officers deemed undesirable. In such cases, established coronéis adjusted to back the new, federally favored state leadership, bending with the political winds in order to retain their status.

Coronel patronage contributed to rural elites' ability to negotiate among themselves over state policies and resource disbursement with little threat of being unseated from below. In the twentieth-century northeast, the voting poor had only marginal influence over political life, limited by their narrow electoral choices. Regional development operated through the reciprocal reinforcement of local, state, and federal government power via patronage networks. Rather than being made obsolete by state programs for economic reform, coronéis became clients of state expansion. Politicians from the northeast used the discourse of retirante suffering to win state patronage for projects that furthered their own ambitions, often with little positive impact on the circumstances of the sertão's landless households and smallholders. The perception in other regions of Brazil that drought aid would be used to benefit wealthy landowners became an obstacle to procuring sufficient federal relief funds for drought victims. Ranchers and estate owners were assumed to have an interest in maintaining the dependent status of their workers to secure their own status as patrons, and thus their pleas for drought assistance were often met with skepticism.[45]

## Drought Aid as Development Failure

At the turn of the twentieth century, Brazilian president Rodrigues Alves created committees to study dams and irrigation in the northeast. Near the end of his presidency Alves established the Superintendência dos Estudos e Obras Contra os Efeitos das Secas (Superintendency of Studies and Works to Combat the Effects of Droughts) to implement the solutions proposed by these committees. In 1909 this became the Inspetoria de Obras Contra as Secas (IOCS; Inspectorate for Works to Combat Droughts) within Brazil's Ministry of Public Works. At the time IOCS was created, the public works minister was the son-in-law of cearense governor Antônio Pinto Nogueira Accioly, who met with the president several times in the months leading up to the inspectorate's creation (and won a third term as Ceará's governor shortly thereafter).[46] IOCS was renamed the Inspetaria Federal de Obras Contra as Secas (IFOCS; Federal Inspectorate for Works to Combat Droughts) in 1919. In 1945 it became the Departamento Nacional de Obras Contra as Secas (DNOCS; National Department for Works to Combat Droughts), which still exists today. Its headquarters were moved from Rio de Janeiro to Fortaleza, the capital of the state most severely impacted by droughts; otherwise, the name changes were of little significance to the agency's administrative organization.

Development economist Albert Hirschman has suggested several reasons why federal aid to the sertão was formally organized in the first decade of the twentieth century.[47] Publication of da Cunha's *Os Sertões* convinced many Brazilian nationalists that the northeastern backlands required more federal attention if Brazil hoped to realize its modernizing ambitions. The U.S. Bureau of Reclamation had been established in 1902 and began undertaking studies to facilitate agricultural cultivation in semiarid Western states. Brazil's civil engineers admired the bureau's goals and early works, and American geologists familiar with these efforts were hired during IOCS's early years to survey the sertão landscape. Additionally, President Alves oversaw sanitation campaigns in Rio de Janeiro to reduce yellow fever and make trade at its port more attractive. The triumph of these hygiene reforms in elites' eyes (despite popular resistance to the urban reorganization and compulsory vaccination that accompanied them), and the international recognition accorded to physician Oswaldo Cruz for his yellow fever campaign, led more Brazilian leaders to embrace science and medicine as essential to national development.

IOCS and its successor agencies have overseen the construction of thousands of dams and reservoirs throughout the sertão. Nevertheless, many late twentieth-century observers like sociologist Renato Duarte felt that the suffering of drought victims had not diminished substantially since the 1910s. DNOCS's primary response to drought was to sponsor "work fronts" for road and dam construction by drought refugees. In 1998 such projects temporarily employed 1.2 million male heads of household (comprising about 10 percent of the drought-affected population) who received minimal rations of beans, biscuits, and brown sugar to feed their families. In the interim between droughts, little was done to prepare for the next disaster. Government response to droughts was crippled by discontinuous funding and by elite opposition to land redistribution as a means of increasing the self-sufficiency and food security of the poorest sertanejos. Agricultural estates continued to be extensively farmed by sharecropping tenants. Commercial farmers focused on marketable oils, other plant extracts, and forage plants for livestock. Few subsistence crops—such as beans, corn, and manioc (cassava)—were grown, except on the family plots of smallholders and sharecroppers.

Ranch productivity did improve with the introduction of hybrid zebu cattle in the 1930s, an initiative sponsored by the federal government. But most people who worked on ranches and farms in the twentieth-century sertão owned little of the land, livestock, and technologies that were the basis of the region's production. Workers supported their families through a combination of subsistence farming, small-scale cultivation for local markets, and labor for their landlords. Sharecroppers were typically in debt to the landowner for seeds, tools, and off-season food, and they generally owed half of their food harvest to him (sometimes in addition to cash rent). They had scant crop surpluses or ability to store them, and they sold their meager goods through intermediaries who took advantage of any price fluctuations for themselves. Sharecroppers' lack of secure work contracts and personal capital made them particularly vulnerable to climatic disasters. Ranch hands were also vulnerable when droughts hit. Following the introduction of improved cattle breeds, they were usually paid in a combination of wages and sharecropping arrangements rather than in livestock (which had historically provided a route to independence, through acquisition of a herd). If cattle populations declined or were moved during droughts, ranch hands were released from work. In contrast, large landowners were relatively cushioned from the economic impact of drought. Farmers with surplus crops could market those at a high price when supplies across the region were scarce. Ranchers moved cattle to less stricken areas during drought years and used them as collateral for loans when needed. Droughts thus exacerbated the social inequities that had begun to increase in the sertão during the 1870s. Many young sertanejos migrated to the south, swelling the infamous *favela* slums of Rio de Janeiro, São Paulo, and other major cities. Yet even with high outmigration, the region's population rose during the twentieth century, putting increased pressure on land and water resources when droughts struck.

Numerous scholars have examined how Brazil's drought agency managed *not* to significantly reduce poor sertanejos' vulnerability to drought during the twentieth century, despite its having a mandate to do precisely that. DNOCS constructed thousands of reservoirs (most of them on private property), but few were linked to irrigation networks to increase agricultural production. Corollary projects like agricultural extension stations, public health posts, and improved education for the poor were pursued only sporadically. As several hydrologic engineers from Ceará pointed out ruefully during conversations I had with them in 2002, the dams that they and their predecessors built were never intended to resolve the drought problem by themselves—they were constructed as the first step in multifaceted development plans. In the absence of other crucial components, water retention had little impact on the plight of the sertanejo poor. With no secure land rights or storage facilities for good harvests, insufficient food reserves, paltry access to credit, and poorly administered relief efforts, smallholders and the landless were repeatedly devastated by droughts.

On the other hand, government aid improved the security of large landholders who had no obligation to sustain their laborers through difficult times—although they were technically required to make water collected in reservoirs built with DNOCS's assistance publicly available during droughts. Stored water offered emergency relief for ranchers' cattle and provided fish for a landowner's family. It was frequently used to generate electricity, a boon to the northeast's industrialists. In Brazilian social critic Darcy Ribeiro's acerbic summary, "A first permanent federal organization—the National Department of Works to Combat Droughts (DNOCS)—created to attend to the problem of droughts has been transformed into an agency of brazen service for the large breeders and the political bosses of the region."[48] Ribeiro cites one indicator of cattlemen's upper hand in much of Brazil's interior: the existence of government-sponsored veterinary facilities for livestock in areas with no equivalent public health clinic for humans. He compares the poorest residents of Brazil's northeastern interior unfavorably to European peasants of past centuries: "No matter how many years or generations he has remained on a piece of land, the sertanejo is always a temporary worker subject to being displaced at any moment without any explanation or rights. Therefore, his home is a hut in which he is only a tenant; his plot is a marginal garden capable only of assuring him the vital minimum to avoid dying of hunger; and his attitude is one of reserve and mistrust, which is fitting for a person living on someone else's land, begging pardon for existing."[49] The humble people Ribeiro describes remain largely dependent upon their landlords, the ranchers and farmers who control essential resources for food production.

Many analysts of this dynamic have argued that the northeast's biggest challenge to equitable development is its landholding pattern. During the twentieth century, farmers who managed to weather droughts—often as a result of development initiatives on their property—increased their holdings by buying out less well-positioned neighbors. According to Renato Duarte's analysis of 1992 landholding data, a mere 7 percent of properties occupied 69 percent of the northeast's rural area in that year, while 75 percent of properties occupied less than 12 percent of the region's rural land (and many workers held legal claim to no property at all).[50] Attempts to redistribute land through federal expropriation of private property, even when the original owner would be paid or receive federally funded irrigation to raise the value and productivity of his remaining property, have repeatedly met with stiff political opposition. Ribeiro asserts that even smaller property owners in the northeast, whose reliance on intensive food cultivation leaves them highly vulnerable to droughts, often make common cause with wealthier landowners in defending the status quo, perhaps fearing the social upheaval that would result if the very poor were to attain some measure of economic independence.[51]

The vulnerability of the sertanejo poor is due to political, social, and environmental factors. Yet through their narrow focus on dam building and other technical infrastructure, the federal government's twentieth-century development agencies attended only superficially to the most marginal sertanejos' predicament without confronting the underlying inequities that make droughts such a catastrophe for them. In the view of many analysts, federal development efforts thwarted more meaningful social change, displacing transformative proposals for resource redistribution while proffering modest paternalistic aid. Contemporary observers still accuse drought aid of undermining rather than enhancing the resilience of the poor by reinforcing debilitating patronage relationships.[52] Some scholars argue that DNOCS and its related federal organizations succeeded in adding a water monopoly to northeastern elites' previous land monopoly.[53]

## Popular Criticism of Sertão Development Efforts

Development plans devised by engineers and politicians for the sertão affected millions of ordinary sertanejos in the twentieth century. Yet it is difficult to know how such people viewed those proposals, or to what extent they were aware of the decisions made by IFOCS and other agencies. Popular responses to drought agencies' work can only be gleaned from fragmentary and sometimes conflicting evidence. One cearense folklorist, Leonardo Mota, captured the rueful humor of his fellow sertanejos with regard to drought assistance; his compilation of popular sayings and anecdotes includes several playful reinventions of the acronym IFOCS that were in circulation at unspecified times during the mid-twentieth century. One is "Isto faz o Ceará secar" (It's this that makes Ceará dry/wither); another is "Isto foi outrora coisa séria" (This was once a serious endeavor); and, finally, "Impossível fazer-se outra cavação semelhante" (It would be impossible to create another equally corrupt business).[54] Clearly the drought administration has been a target of popular ridicule at various points in its history.

One window into sertanejo reaction to droughts and the government's effort to mitigate their impact is a form of northeastern folk poetry known as the *cordel*. *Cordéis* have been sung and sold in regional markets since the brief economic expansion caused by the 1860s cotton boom. They are printed as small pamphlets (*folhetos*) of about eight pages containing one or more multistanza poems. The *folhetos* have traditionally been published by the poets themselves, using hand-cranked block presses, though in recent decades prolific authors print from personal computers. According to Candace Slater, a scholar of Nordestino popular literature, most cordéis are printed in runs of between ten thousand and twenty thousand copies, with two thousand being about the lowest circulation. Some popular titles run to 100,000 in multiple editions.[55]

Slater refers to traditional cordel authors as "culture brokers" who communicate current events to their less traveled listeners and readers. The importance of this role has diminished with sertanejos' increasing access to mass media, but cordéis continue to provide a rich source of popular reflection on political and social life in the northeastern backlands. Many authors and consumers of cordéis understand the verses to have an inspired moral message and to authoritatively express the sertanejo experience. Of course, authors' opinions vary, so the poems do not present a consistent narrative or analysis when taken as a whole. Renato Campos, whose literary research focuses on *cordéis*, identifies no uniform political ideology shared by cordel authors, and he observes that they rarely recommend specific changes, even when their poems highlight particular social or political problems.[56] Folklorist Mark Curran, who interviewed one of the most prominent northeastern *cordelistas* over a period of several decades, describes cordéis as depicting good and evil forces at work in the lives of the poor without embracing a clear political position.[57] The poets act as trusted chroniclers and moralists with regard to current events that affect the disenfranchised.

Slater posits that although cordel authors do not generally confront injustices forthrightly in their writing, they do tend to sympathize with the poor in their ongoing negotiations with the wealthy and powerful. As one anonymous poet told her, "I feel sorry for the poor man, and even if I were to become rich, I would still like poor people better. Me, I write for the Christs. Do you know what a Christ is? It is someone who works for another person. Well, then, I write for him, and the only reason I don't talk more about his sufferings is that the others, the bosses, would kill me if I did."[58] As this statement indicates, cordel poets often cannot risk the retribution that they might suffer if they aggressively criticized an individual coronel or unjust state institutions. Alliance with wealthy patrons is understood by most *cordelistas* to be a necessary survival strategy for the northeast's poor. Patrons who protect their clients are typically praised for fulfilling the role that life has handed to them as benefactors of the less fortunate.

Numerous cordéis dealing with drought ask the government to provide more aid to the sertanejo "martyrs."[59] One satirical example, written in the year that IOCS was formed, depicts a personified "tax" and "hunger" in conversation about the protections that they enjoy from the government. Hunger is fearful that the president will destroy him with the new projects to alleviate drought, but tax is confident that no harm will come to them. The poet Leandro Gomes de Barros lamented anarchy and injustice in the sertão and criticized the imposition of taxes from which the public received no benefit. He observed that the loss of vegetation from repeated droughts was placing the sertão under further privation, which the government did little to mitigate.[60] A native of Paraíba state who later moved to Pernambuco, Gomes de Barros wrote another cordel a decade later about government mismanagement of drought aid; "A Sêcca do Ceará" jokes that the money sent from Rio de Janeiro for relief is shrewd and likes to hide in the coffers of powerful people, finding the pockets of drought victims (flagelados) to be unacceptable lodging. The federal government had responded to the need for drought aid by raising taxes due from states, but this policy was killing Nordestinos, Gomes de Barros argued, because states met their federal obligations by taxing everything in sight.[61] This established poet's overt criticism of state policies is unusual in light of folklorists' assertion that cordel authors were hesitant to oppose authority forcefully. In several of the chapters that follow, the lyrics of folk poets and popular musicians provide some insight into ordinary sertanejos' views of drought aid and state development schemes.

### Drought as a Natural and Social Phenomenon

In 1900, Euclides da Cunha published three articles in the newspaper *O Estado de São Paulo* recommending ways to confront drought in the northeast.[62] He portrayed the sertão as a potential national asset, salvageable through engineering. Rejecting a theoretical link between the drought cycle and sunspots, da Cunha concluded that a series of geographic features combined to produce droughts, and he called for systematic observations of the variety of factors influencing the northeast's climate in order to understand the multiple causes of drought. The engineer compared the northeast unfavorably to other semiarid regions like French Tunisia, where drought- and flood-control mechanisms implemented by the ancient Romans were still the basis of irrigated agriculture at the turn of the twentieth century. Da Cunha believed that in the sertão the absence of flood control and irrigation had contributed to desertification; he advocated creating numerous small reservoirs throughout the region's river valleys, as had been proposed since the Great Drought. Such a network of lakes would alter the sertão's weather through increased evaporation and reduce the frequency and severity of droughts, da Cunha believed.

This turn-of-the-century analysis describes drought as a climatic phenomenon—which, of course, it is. But in northeastern Brazil, the crisis caused by drought is also "a socio-economic [phenomenon] related to the acute vulnerability of the rural population."[63] Recent historiography of natural disasters, influenced by Amartya Sen's work since the 1960s, emphasizes that the impact of such events on human populations depends on the political and economic organization of the societies affected.[64] People who subsist on the margins of the political or economic order prevailing in their locales are most likely to suffer when hurricanes, droughts, or famines strike, while those in more secure positions are better able to weather such calamities. Yet the administrators who oversaw drought works in Brazil's sertão during the early twentieth century rarely discussed the social inequities that made drought a catastrophe for the sertanejo poor; they emphasized instead their agency's efforts to improve the region's outdated hydrologic, transportation and agricultural infrastructure. Sertão developers' emphasis on the technological and climatic problems that contributed to drought crises overlooked the imbalance of power among social sectors that left millions of landless sertanejo families and small farmers particularly vulnerable to starvation.

· · · · · ·

This chapter has examined how the sertão and its population were understood in national debates about Brazilian modernization during the late nineteenth and early twentieth century, and it indicates the intersection of natural and social factors that conspired to marginalize the sertão's landless poor in negotiations for state and environmental resources. Many Brazilians assumed that the racially dubious, culturally backward, and often illiterate sertanejo population could not contribute productively to the national economy. Sertanejos seemed foreign to many of their coastal compatriots, and this attitude made it easier to overlook their recurrent suffering. Pleas for assistance made on their behalf were often met with little sympathy.

Nonetheless, from the 1870s on, northeastern elites framed demands for national investment in their region as a humanitarian appeal to reduce the suffering of drought migrants. By explaining poor sertanejos' precarious existence as a problem of insufficient technical infrastructure, they deflected attention from the inequities that left landless families and small farmers vulnerable to myriad misfortunes. Elites' emphasis on drought as the sertão's central problem and public works as the appropriate solution designated civil engineers as the key agents of regional transformation. Engineers, due to their training and social position, generally embraced a technocratic development ideology requiring little disturbance of—or even attention to—the social order.

During Brazil's First Republic and the decades that followed, varying interpretations of the sertanejos' plight—as rooted in social or environmental causes—suggested different ways to understand and ameliorate their misfortune. The range of development proposals that emerged reflect a spectrum of views on the sertão's fundamental challenges; they were often based on prevailing national stereotypes about sertanejos' racial character and their potential as citizens of a modern nation. The following chapters analyze political and technological solutions to the drought problem pursued from 1909 to 1964 by public health reformers, engineers, agronomists, economists, and politicians. They consider to what degree these historical actors believed that scientific evaluation and technical aid could transform the sertão and what this says about their vision for twentieth-century Brazil—including sertanejos' place within it. The chapters also assess which interpretations of the sertão's periodic crisis held the most promise for improving the general welfare of sertanejos and why such agendas were or were not embraced by those with the power to carry them out.

# 2 Civilizing the Sertão

## Public Health in Brazil's Hinterland, 1910s

· · · · · · · · · · · · · · · · · · · · · · · · · · · · · · · · · · · · · · · · · · · · · · · · · · ·

Brazil's federal government began surveying the health of its rural population during the early twentieth century, and these surveys became the basis of a wide-ranging debate about the root of "backwardness" in the
country's interior. In the northeast's semiarid sertão, health surveys were
undertaken in preparation for the infrastructural modernization planned
by the new drought agency.

This chapter explores two contrasting interpretations of the sertão's economic lethargy, both of which were offered by public health campaigners.
One perspective, emphasizing the economic and environmental hardships
faced by the region's residents, known as sertanejos, was articulated most
actively from the 1910s through the 1920s by physician Belisário Penna, who
was among a new group of Brazilian sanitation enthusiasts who rejected
pessimistic racial determinism due to their faith in the redemptive power
of modern science. A contrasting perspective, dismissive of sertanejos' potential as contributors to modern Brazil, is represented here in assessments
made by Americans from the Rockefeller Foundation's International Health
Board (IHB), and particularly its director Wickliffe Rose. Views similar to
Rose's were expressed by some Brazilian doctors as well, in negative reports
about sertanejos' modernizing potential. Understanding the contours of
public health debates regarding the sertão and its inhabitants during the
1910s and 1920s facilitates a critical reading of drought engineers' sertão
development discourse during those same decades, the subject of chapter 3.

Penna became the first director of Brazil's Serviço de Profilaxia Rural
(Rural Disease Prevention Service) in 1918, after surveying health and hygiene in several interior regions. Through this experience Penna came to
see the rural poor as oppressed by debilitating natural and social environments. Rural Brazilians at the turn of the twentieth century were exposed
to unsanitary living conditions and rampant disease. Many endured near-
servile conditions as sharecroppers on vast estates. The sertanejos of the
Nordeste (northeast) region faced particular hardships as a result of periodic droughts and extreme isolation from the rest of the country. In articles,

books, and speeches written from 1916 to 1930, Penna demanded that politicians confront rural Brazil's misery with substantial federal resources. For backlanders to contribute to national progress, he argued, the republic's government had to invest in public health throughout the interior as well as drought alleviation in the northeast; it also needed to reduce the tyranny of regional oligarchs by redistributing land to agricultural workers.

Penna viewed aid to rural populations as a moral alternative to the importation of agricultural laborers from Europe that Brazil's southern coffee planters had embraced since the 1880s. He saw the enervation of Brazil's rural poor as a reflection of the republic's political failings. In contrast, the IHB's Rose accepted racial explanations of rural Brazilians' low productivity; he viewed race, rather than politicoeconomic organization, as the most significant feature distinguishing different regional populations within Brazil. From this Rose concluded that aiding regions with the largest proportion of people descended from Europeans—notably the state of São Paulo—was the surest way for the IHB to promote economic progress. Rose and his colleagues imported an American racial ideology that assumed an absolute categorical distinction between blacks and whites. This placed them at odds with Brazilian intellectuals who interpreted race more fluidly, as a spectrum of differences that varied as a result of intermingling among African, European, native American, and Asian peoples.

Penna's and Rose's differing attitudes toward rural populations stemmed in part from their respective degrees of interaction with them. From his extensive travels and interior health surveys, Penna had firsthand experience of the ways in which rural laborers were neglected by government authorities. As he and his collaborator Arthur Neiva wrote in their 1916 report on health conditions in the northeastern sertão, "We still retain vividly the sorrowful impressions of profound misery and abandonment in which thousands of human beings lie, and our testimony should in some form work to mitigate their suffering."[1] Rose, on the other hand, had no personal encounters with rural Brazilians. He was primarily concerned with the "demonstration effect" of the IHB's Latin American public health programs, which would be more dramatic in areas with more responsive governments (such as São Paulo).

Penna and the IHB staff also differed in their interpretations of the root causes of disease. The IHB viewed infectious diseases as purely biological entities. Penna, on the other hand, saw rampant disease among the rural poor as the result of a retrograde political and economic system that made workers vulnerable to infection. He thus believed that public health efforts were most sorely needed in areas with negligent local governments. Through his lectures, publications, and work for the federal health department, Penna hoped to alter the relationship between Brazil's vast hinterland and its political and economic centers, better integrating the nation's expansive land mass and diverse population. He invoked nationalist themes employed by numerous reformers during the First Republic who tried to inspire the federal government to pursue progressive modernization throughout the country.

The contrast between Penna's assessment of sertanejos' potential and those of IHB staff in Brazil during the 1910s provides an instructive delineation of the variant ideologies underlying scientifically based regional development in this period. As will be argued in chapter 3, engineers working for the federal drought agency in the sertão frequently found themselves caught between these ideological poles as they grappled with the potential and limits of science and technology as tools for regional reform.

### The Sertão's Skeletal Health Infrastructure

Until the federal government's Rural Sanitation Service was established in northeastern Brazil, in 1920, states in that region provided almost no health services outside of their capitals. This is unsurprising given limited budget allocations for public health and the paltry votes that expenditure on the interior could garner for a governing party.[2] When physician Fred Soper of the IHB surveyed Pernambuco's hygiene services in 1920, the state employed twenty-four health inspectors, twenty of them in the capital, Recife. Only 10 percent of the state had clean drinking water, and Recife was the only town with a sewage system.[3] Soper found that, except during epidemics, health problems in the Pernambucan hinterland were generally ignored.[4] Reports by the hygiene inspectors of several northeastern states confirm the dearth of public health investment in the interior. Droughts and the epidemics that accompanied them, due to malnutrition and mass migration, posed significant challenges for the sertão's highly inadequate health administration.

Data from the Pernambuco state hygiene inspector's annual reports during the 1910s indicate how limited health services were in the sertão. Of Pernambuco's 175 hygiene inspectorate employees in 1912 (not including two hundred *guardas* who comprised a mosquito extermination brigade to reduce yellow fever incidence), only ten were assigned to the Serviço do Interior: five inspectors and five assistants.[5] Of the vaccines distributed by the inspectorate during that year, 70 percent were used in Recife.[6] With weak municipal governance and a state preoccupied by the problems of its capital, the health inspector had no means of addressing disease outbreaks in the sertão. Railway lines were a major source of disease transmission between the coast and the interior, yet municipalities along the lines neglected to enact any sort of public health measures.[7]

Northeastern capitals were besieged during drought years by migrants who gathered in makeshift camps on the cities' outskirts; refugees flocked to the capitals because it was there that they had the greatest hope of receiving assistance. Recife's hospitals and philanthropic organizations were overwhelmed by an influx of *retirantes* (drought migrants) from their own and neighboring states' *sertões* during the droughts of 1915 and 1919. The governor's annual report for 1915 describes the waves of desperate migrants as if they were immigrants from other lands, even though many were from Pernambuco but had no access to public assistance in their home municipalities: "As in 1914, there occurred a noteworthy immigration of sick and infirm people of all sorts, originating from the interior of our state and neighboring ones, who, flogged by pressing needs and after completely exhausting the philanthropic potential of the sertanejo population (which had used up all of its resources and economic means as a result of the intense and prolonged drought that sterilized the soil), came to emigrate to Recife's hospitals, in order to obtain temporary or final rest for their bodies from ills, most of which had no cure."[8] Of 13,913 people registered as patients in Recife's hospitals that year, 62 percent were "imports" from the interior of Pernambuco or nearby northeastern states. More than half of the city's 3,484 deaths were of drought retirantes, one-third of whom died of tuberculosis. A smaller but significant fraction succumbed to dysentery caused by contaminated water. Epidemics of smallpox and yellow fever also raged among the famished refugee population, most of which had not been exposed to those diseases previously.

During the 1919 drought, Pernambuco's governor José Cavalcanti accused his state's rural municipalities of being too dependent on Recife in times of crisis, noting, "Due to the absence of hospital facilities in our interior cities, where they ought to exist because many of these have large populations and abound with public resources and philanthropies along with profuse rail networks for communication, their indigent are attracted by our hospital services, aggravating and weighing upon the death rate in Recife."[9] These concerns prompted Cavalcanti to invite the IHB to establish rural hygiene posts in Pernambuco in 1920 under a cooperative agreement with the federal government.

The governors' annual reports detail the crippling effect of droughts on food production and the consequent malnutrition of the rural and urban poor. In 1915, decreased agricultural output led to a rise in the price of basic foodstuffs, putting Recife's workers at grave risk. In response to pleas for assistance from interior municipalities, the governor sent seeds, which would only have been helpful if rain had arrived.[10] During the 1919 drought, inflated food prices due to the missed harvest and diminished cattle population again posed a threat to public health, yet the reduction in revenue from agricultural exports led the government to reduce public expenditures.[11] Limited state outlays for public welfare became even more restricted during droughts, with the justification that the agricultural economy would be paralyzed if taxes were increased to address these emergencies.

A history of public health in Paraíba, one of two small states along the northeast coast between Pernambuco and Ceará, provides further confirmation of minimal public health administration in the region prior to 1920. In the early 1890s, Paraíba's public health service consisted of only one inspector who registered sanitation complaints with municipal and state officials but had no power to enforce compliance with sanitary codes. A federal regulation in 1911 required that states be divided into public hygiene districts and specified municipal and state responsibilities, but this had little impact anywhere except in Paraíba's capital. In 1912 plague erupted in Campina Grande, the state's second largest city—located on the border between Paraíba's semiarid sertão and its more fertile *agreste*. No representative of the state hygiene service worked in the city, and although the director of hygiene received a limited supply of vaccine serum from Rio de Janeiro, he did not know how to use it. An association of Campina Grande cotton merchants finally contracted a renowned doctor from Recife, Octavio de Freitas, to develop a strategy for dealing with the epidemic, since their state government was evidently powerless to help.[12]

In her detailed study of Paraíban politics during the First Republic, Linda Lewin notes the "drastic contrast in living standards" by 1920 between northeastern states and those of the industrializing south that impacted both public health and political incentives to invest in it. In 1922 more than 80 percent of houses in Paraíba's capital were constructed of mud and straw thatch (rather than brick and tile), and only 5 percent of homes were connected to a sewage system. The infant mortality rate in the city was 50 percent higher than in Rio de Janeiro, at 217 versus 154 per thousand children. The state's literacy rate remained a paltry 13 percent—half that of Brazil as a whole, according to the 1920 census—and this limited the number of men who had a voice in political affairs.[13] Data for Paraíba's sertão region is neither reliable nor readily available for this period, but it is reasonable to assume that infant mortality, poverty, and illiteracy were significantly higher there than in the capital.

Ceará, the state most impacted by drought since its semiarid zone comprises roughly 90 percent of its territory, displayed similar incompetence in combatting infectious disease during the first decades of the twentieth century. Although *cearense* colonial administrators advocated the use of the vaccine against smallpox as early as 1806, no adequate vaccination program was in effect even a century later.[14] The position of state hygiene inspector was finally established in 1893, and within two years the inspector had plans to tabulate mortality statistics, begin vaccine production, and provide bacteriological analysis. These services were intended primarily for the capital, Fortaleza, although the hygiene inspectorate could help to fund services organized in the interior.[15] But subsequent annual reports indicate Ceará's lack of commitment to public health. In 1897 the inspectorate still had no designated building, and its activities had to be carried out in the inspector's own medical office. Inspectors frequently recommended that a state vaccine institute be established, but they relied for many years on smallpox vaccine imported from Rio de Janeiro. Fortaleza's unchanneled water and sewage offered no straightforward way to block transmission of diseases carried by water or human waste. The city experienced various epidemic outbreaks, including one of yellow fever in 1912.

The depth of Ceará authorities' negligence with regard to public health is illustrated by the biography of legendary pharmacist Rodolfo Teófilo, son of a Bahia medical faculty graduate who was employed by Ceará's government in the low-paid position of "doctor for the poor." As a child in the 1850s, Teófilo witnessed yellow fever and cholera epidemics, the latter of which claimed his newborn sister. When he was eleven his father died, succumbing to depression after falling into bankruptcy. Teófilo spent his late teenage years apprenticed to a cotton merchant, who treated him as a servant. He won a fellowship to study pharmacy in Bahia and returned to Ceará at age 24, just as the Great Drought struck in 1877. Teófilo's early experience with illness and poverty, and his witnessing of how roughly the police dealt with drought migrants—whom they viewed as a threat to public security—shaped his medical career.

Teófilo spent the years following the Great Drought treating the orphans of drought migrants in Fortaleza. When another drought struck in 1900, Teófilo was in Salvador, Bahia, where he obtained smallpox vaccination equipment and training from the Instituto Vacinogênico; he then returned to Ceará. The governor had acquired some vaccine from Rio de Janeiro, but it was largely ineffective, having lost potency in transit. Exasperated, Teófilo began manufacturing vaccine in his own home. He inoculated calves with serum sent by a colleague in São Paulo and used lymph from the pustules that formed on their skin. After testing his vaccine on prominent citizen volunteers, Teófilo bribed wary inhabitants of the city's slums to be vaccinated. Failing to persuade the government to make vaccination obligatory, he offered free supplies and instruction to doctors in interior municipalities who were willing to join his statewide campaign. Teófilo has been credited with eliminating smallpox from Ceará within two years. The epidemic reappeared a decade later, carried along railroad lines, and Teófilo resumed his private vaccination campaign.[16]

Teófilo made it his mission to publicize the suffering of cearenses during droughts. Among his efforts are a series of poems about sertanejo life that include several heartbreaking descriptions of the drought exodus. In one, a man who has lost faith in his farm and his patron saints decides to leave his family to seek wealth in the Amazon, where his wife fears he will die of fever.[17] Other poems offer gruesome accounts of children and parents in agony as they watch each other die from hunger, and of the madness that accompanies starvation in humans and animals. Teófilo's writings emphasize the humiliation of having to abandon a home and land that had provided sustenance for years in order to beg for charity. His numerous books describing sertanejos' misery during droughts chastise the state for neglecting them. An account of the 1915 drought begins, "Ceará is a land condemned more by the tyranny of its government than by inclement Nature."[18] The pharmacist called upon the press to chronicle the suffering of drought victims and garner public sympathy, just as some newspapers had aided the abolitionist cause in the 1870s by portraying slaveholders' abuses.

Teófilo's history of the Great Drought (1877–79), which includes photos of famished people and the plants that they relied on to escape starvation, won him election to the prestigious Instituto Histórico-Geográfico Brasileiro, founded in 1838 as Brazil's preeminent scholarly society. In it he explained his motivation for repeatedly drawing attention to Ceará's drought victims "as a scream of alarm to future generations, and also as a protest based in facts, all authenticated, against the indifference of the public officials to the suffering, the miserable conditions of the region assaulted by droughts. In the south of Brazil they are unfamiliar with these calamities and they estimate the value and energy of cearenses to be insignificant; this unhappy and captively martyred people whom one sees time after time in open struggle with the most relentless misfortunes."[19] Among the policies that Teófilo proposed to alleviate the suffering of drought victims were tax incentives to encourage manioc cultivation as a reserve food crop. Teófilo believed that dams and irrigation works could also diminish the disastrous effects of climate instability and increase agricultural production in Ceará; he chided the federal government for its paltry expenditures on drought relief and insufficient construction of reservoirs.[20]

In 1915 Ceará's hygiene inspector described in graphic detail the health impact of mass migration from the sertão to Fortaleza.[21] Teófilo's vaccination campaign had preempted a devastating smallpox epidemic, and vaccines continued to be distributed—about half of which were manufactured by Teófilo, another quarter sent from Rio, and the rest made by doctors employed by the state. Even so, the death rate in Fortaleza rose by as much as 900 percent per month, as insects—feeding on the unmanaged waste of seventy thousand migrants gathered in a disorganized camp on the city's outskirts—began spreading diseases to urban residents. In the camp, which inhabitants referred to tellingly as a "corral,"[22] there were only nine doctors available to treat the thousands of sick. Young children died in large numbers from drinking infected water and milk. The health inspector estimated that 300,000 people were dislocated by the drought throughout Ceará, and perhaps between thirty thousand and sixty thousand of these died, many of them on their way home from encampments where they had received insufficient medical attention and unclean food.

In response to these horrors, the inspector proposed intensified sanitary measures in Ceará's interior, including digging deep wells from which uncontaminated water could be obtained. Frustrated with his lack of influence over health conditions in the sertão, physician Costa Ribeiro wrote, "Hygiene delegates—mere decorative posts for political cronies in the interior, which professionals frequently recuse themselves from accepting—have not been affected by reform to a degree that would assure us, at the very least, of receiving information about what is happening in the state [with regard to disease]."[23] Epidemics remained difficult to combat in part because health and demographic statistics were scarce and unreliable outside of the capital. Yet despite this abysmal level of public administration, the state's hygiene staff was reduced as soon as the drought ended. Four years later, when another drought occurred, Ceará's health service (elevated in August 1918 to Diretoria Geral de Higiene—general directorate of hygiene) employed only sixteen people. The sertão remained severely underserved, without adequate access to vaccines or sanitary infrastructure.

As these accounts from three northeastern states indicate, sertanejos were often viewed by their state governments as undesirable foreigners when they sought help in capital cities. Although health and sanitation services were woefully inadequate in the interior, drought retirantes' claims on medical assistance in the capitals (when they managed to travel that far) were resented, and their mass presence inspired fear in urban residents. Even Teófilo, whom many regarded as a hero for his selfless labors on behalf of drought victims, was thwarted by the oligarchy because he publicly condemned Governor Antônio Pinto Nogueira Accioly for paying scant attention to health problems, particularly in the sertão.[24] Such neglect of sertanejos by northeastern state governments led sanitary reformers like Penna to believe that the health needs of rural Brazilians must become the focus of *federal* aid in the interest of national progress.

## Embracing Sertanejos as Citizens

Penna was born in 1868 to an established family in Barbacena, Minas Gerais, a politically influential state. Following his education at the medical schools of Rio de Janeiro and Salvador he embarked on a clinical career in Minas Gerais during the 1890s. A decade later he joined President Rodrigues Alves's sanitation campaign in Rio de Janeiro as a sanitary inspector under the renowned bacteriologist Oswaldo Cruz. In 1905 Penna was transferred to the prestigious Serviço de Profilaxia da Febre Amarela (Yellow Fever Prevention Service), where his recommendation of weekly home inspections to remove *Aedes aegypti* mosquito larvae increased the effectiveness of Cruz's campaign to eradicate the disease. In 1907 Cruz sent Penna and Carlos Chagas to combat malaria along the Central do Brasil railway in the north of Minas Gerais; Penna remained on this mission through 1910. Cruz then invited Penna to accompany him in combating malaria along the Madeira e Mamoré rail line in the Amazon basin, and Penna subsequently spent a year organizing a yellow fever campaign in the Amazonian state of Pará. He thus had substantial exposure to living and sanitary conditions across Brazil during the republic's first two decades.[25]

In 1912 Penna embarked on a six-month sanitary survey of the northeastern backlands, accompanied by physician Arthur Neiva of the Instituto Oswaldo Cruz—another veteran of Cruz's Yellow Fever Prevention Service.

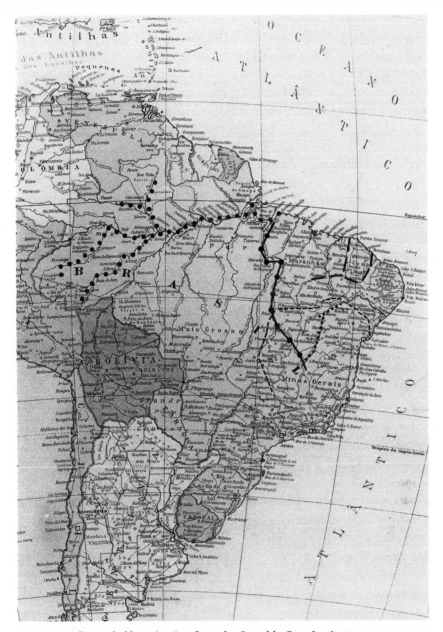

Sertão expeditions led by scientists from the Oswaldo Cruz Institute, 1911–13. Belisário Penna and Arthur Neiva's route is marked by small dashes originating at Salvador. Source: Thielen, *Science Heading for the Backwoods: Images of the Expeditions Conducted by the Oswaldo Cruz Institute Scientists to the Brazilian Hinterland, 1911–1913* (Rio de Janeiro: Instituto Oswaldo Cruz, 1991), 11.

Neiva had just returned from a two-year research trip to the United States, where he was impressed by the nation's public health administration and scientific education. Penna and Neiva's trip was sponsored by the newly formed Inspetoria de Obras Contra as Secas (IOCS; Inspectorate for Works to Combat Droughts) to assess health conditions in the sertão prior to initiating dam construction there. IOCS funded two other expeditions as well. Physicians João Pedro de Albuquerque and José Gomes Faria traveled to Ceará and Piauí but left no textual record of their observations. (A photographic record remains, held at the Casa de Oswaldo Cruz in Rio de Janeiro.) Physicians Adolpho Lutz and Astrogildo Machado, who surveyed the São Francisco River valley, published a report emphasizing the population's ignorance and racial degeneracy.[26] Penna and Neiva, in contrast, explained sertanejos' backwardness as arising from cultural isolation and government negligence rather than immutable racial handicaps. Their report, published in 1918, became the most influential analysis of rural health in this period; it described sertanejos as "abandoned" by their government to primitive conditions and burdened by numerous illnesses, all of which reduced them to a miserable existence. Penna and Neiva's rejection of fatalistic arguments about Brazil's racial composition and climate brought their analysis considerable attention at a time when the new republic's prospects as a modern nation were widely debated by intellectuals and politicians.

Penna had likely encountered theories of disease causation adapted to Brazil's particular national circumstances during his year at the medical school in Salvador, Bahia. Doctors there had been arguing since the 1860s that many diseases attributed to tropical climates (by environmental determinists, following a European colonial medical tradition) were instead caused by the social conditions frequently found in such regions, including poor hygiene and diet—the legacies of colonialism and slaveholding.[27] Bahia's *tropicalista* physicians embraced a neo-Lamarckian interpretation of environmental influence on human populations, believing that social conditions could influence heritable traits, for better or worse. Penna's frequent assertion that the living conditions of Brazil's rural poor must be altered in order to improve their health and productivity reflected these views. Such ideas were reiterated by Afranio Peixoto, professor of hygiene at Rio de Janeiro's medical faculty, in the early twentieth century. Peixoto also opposed environmentally and racially deterministic interpretations of disease in Brazil and rejected European-style "tropical medicine" because it presumed that diseases prevalent in the tropics thrived there primarily as a result of the climate.[28] He praised Penna and other sanitarians for their empirical approach to combating Brazil's endemic ills and their emphasis on political and social conditions as the root causes of disease among the poor.[29]

Penna and Neiva's analysis of cultural backwardness and low productivity in the sertão exemplifies a shift in Brazilian racial discourse that occurred during the early decades of the republic. In the late nineteenth century, many Brazilian intellectual and political leaders were troubled by European racial theories that placed groups along an evolutionary hierarchy, with whites at the top. They rejected the climatic and racial determinism espoused by Europeans engaged in colonial expansion, who assumed that tropical climates were less salubrious than temperate ones and therefore permanently enervated their inhabitants. According to those dual logics, Brazil, with its racially mixed population of native (índio) and African descent living largely within the tropics, was doomed in its modernizing ambitions.[30]

Brazil's republican elites countered this prophesy with an embrace of racial mixing suited to their national circumstances. São Paulo's coffee planters, in particular, became avid supporters of *branqueamento* (whitening) policies intended to alter the racial makeup of Brazil within a few generations while adding to its free labor force. The basic premise of branqueamento ideology was that when whites married people from theoretically weaker races, their children would exhibit a predominance of superior, European traits, slightly diminished by the influence of the less advanced races. Over time, the union of whites with mixed-race Brazilians would "breed out" less desirable characteristics of Africans (especially) and índios, producing a fully white and culturally European population. This notion was put into practice by *paulista* planters, who cooperated with the national government in the 1880s (once slave emancipation was clearly inevitable) to subsidize the passage of European immigrants to work their coffee farms.[31] Since many European laborers refused to toil alongside slaves, the perceived desirability of importing them became one motivation for finally abolishing slavery in 1888. Abolition, in turn, justified greater immigrant recruitment to replace the slave labor force, because southern planters viewed freed slaves as degenerate and unmotivated when compared to European farmers.[32] Rather than offering jobs to freedmen following abolition, coffee producers hired German, Italian, and Japanese families to work their estates whenever possible.[33]

Like their contemporaries among Brazil's elite and professional classes, Penna and Neiva believed in a hierarchy of races. This is evident in the disdain expressed in their northeastern rural health survey for supposedly African traits exhibited by the population and in their recommendation that European immigrants be encouraged to settle the sertão and model industrious farm labor. Yet they differed from other critics of Brazil's African and indigenous influences in that they did not propose eliminating traits attributed to non-European ancestry through interbreeding with whites. They suggested instead that the Northeast's mixed-race, rural poor would progress if guided by the *example* of vigorous and "cultivated" white settlers (*povos cultos*) and provided with technological infrastructure linking them to the civilized world.[34] This reflects their Lamarckian faith in the possibility of improving a racial stock through education and technology.[35] Penna and Neiva insisted that European immigrants (who had not been encouraged to settle the northeast because of its supposedly inferior soils) could make a reasonable living along the perennial São Francisco River if the government constructed adequate transportation and communication networks to the coast. They therefore proposed giving the best land in the semiarid sertão to foreigners so that the immigrants' eventual prosperity would inspire the existing population to greater effort and thereby fuel economic growth throughout the region.[36]

Penna and Neiva's insistence that Brazilian modernization must be adapted to the country's complex racial heritage was echoed by many nationalist writers during the early twentieth century. Several prominent intellectuals were more optimistic about racial intermixing than dominant European or North American ideology of the time was. Manoel Bomfim, who also attended Bahia's medical school in the late nineteenth century, published a widely read essay on Latin American history in 1903, in which he questioned the evidence for the inferiority of mixed-blood *mestiços* and the significance of racial differences. He believed that adherence to discredited European racial theories like polygeny (the belief, inconsistent with Darwinian evolution, that each race has a separate historical lineage) had led Latin Americans to neglect their own rural populations in favor of European immigrants. Latin American nations should focus their resources on popular education, he argued, for the improvement of native inhabitants who had been neglected under colonial rule and decades of virtual (or actual) enslavement.[37]

Alberto Torres, an important figure in the Brazilian republic's early years, published numerous articles in newspapers in Rio de Janeiro and São Paulo following his retirement from the country's highest court in 1909;[38] in these he encouraged elites to pursue development strategies adapted to Brazil's particular circumstances rather than adopting models from foreign countries. Torres opposed European immigration and branqueamento ideology, rejecting assumptions of racial hierarchy and the notion that racial composition affects a nation's economic potential (although he was dubious about racial mixing, which he thought might be degenerative).[39] Torres argued that Brazil should industrialize slowly, exporting a few agricultural products like coffee in return for foreign manufactured goods, and he believed that greater knowledge of Brazil's natural resources and a more equitable distribution of land would allow the country to both feed its population and prosper. Both Penna's and Torres's writings exhibit nostalgia for the largely rural economy of Brazil's nineteenth-century empire and the more centralized authority of that imperial system; both asserted, paternalistically, that Brazil was not ready for democracy due to the very limited education (civic and otherwise) of its citizenry.[40] They believed that responsibility for guiding the young nation along a path to modernization lay with politicians and professionals like themselves. Many of Torres's arguments reappear in Penna's published work, including opposition to urbanization and the government investments that encouraged it; suspicion that industrialization would be harmful; promotion of a strong central state to protect individuals' rights in far-flung regions; and the assertion that many Brazilians were debilitated by poor hygiene, malnutrition, and insufficient education.

Torres expressed a dissatisfaction with Brazil's republican government that many Brazilian intellectuals, including Penna, had come to feel by the 1910s. The republic's 1891 constitution shifted Brazil from a centralized administrative structure to a decentralized federalist association of states. This paved the way for oligarchic usurpation of state power. Middle-class supporters of the republic's formation soon contended that the new nation was a mere conglomeration of fiefdoms with no ambition to unify. Since states could act as nearly autonomous units in fiscal and legislative matters, existing regional differences became more pronounced. By the time Penna and Neiva published their northeastern rural health survey in 1918,[41] many Brazilians were frustrated by the decadence of a republic whose policies benefited oligarchs but overlooked the crowding and disease of urban slums and the myriad needs of rural areas. The brutality of World War I increased Brazilians' receptivity to the idea that they should free their nation from imitation of debased European or North American institutions.

Rural sanitarians continued to argue into the 1920s that Brazil's hinterland population was burdened by curable diseases, not immutable racial handicaps. This emphasis on environment as more influential than hereditary racial endowment became a central tenet of progressive and "positive" eugenics movements in urban as well as rural Brazil through the 1930s. Penna and Neiva's advocacy of greater government investment in Brazil's countryside, particularly in the northeast, during this period placed them at the forefront of a nationalist movement that gained adherents during the 1920s, ultimately bringing the flawed republican experiment to a close with the revolution led by Getúlio Vargas in 1930.

## Penna and Neiva's Sertão Health Survey

Belisário Penna and Arthur Neiva began their sertão odyssey in March 1912 with a three-day steamship journey from Rio de Janeiro to Salvador, Bahia. They then boarded a train for the five-hundred-kilometer trip to Joazeiro, an interior town of six thousand inhabitants on the bank of the São Francisco River near the convergence of the states of Bahia, Pernambuco, and Piauí. Their train carried water for distribution to parched sertão communities. The doctors found Joazeiro "primitive," lacking paved roads, electricity, or a sewage system. Water was not piped into homes; instead, mules transported it from the river. Malaria and typhoid were rampant and, despite the scarcity of water, some interior hamlets near lakes and along rivers had been depopulated to escape mosquito-borne illness.

After two weeks of preparation, the physicians' caravan of ten people and twenty-four burros crossed the kilometer-wide river to Petrolina, where they began their trek into the sertão proper. The following ten weeks would take them across five hundred kilometers of semiarid terrain, mainly in Piauí along its border with Bahia. The company progressed an average of four leagues (under fourteen miles) each day, first westward across the sertão then southward toward Goiânia. Finding clean water for humans and pack animals to drink was a constant preoccupation; at times the distance between water sources was over ten leagues (nearly forty miles), or three days' travel. Since water evaporated rapidly in the dry heat, information about where potable water could next be obtained had to be regularly updated from the occasional traveler heading in the opposite direction. Little contact with the outside world was possible via newspapers or post, although the entourage passed a few scattered telegraph stations. The expeditionaries went months without hearing from their families.

In their published report recounting this journey, Penna and Neiva describe sertanejos as "indolent," leading "vegetative" lives with minimal comforts. Markets and businesses of any kind were scarce, and money was rarely used; essential transactions were based on an exchange of goods. The sertão seemed "impenetrable to progress . . . primitive," with no sign of the surrounding world's industrial advancements.[42] Sugar mill technology was less efficient than it had been under Dutch colonial rule three centuries before, the doctors asserted. Even coffee grinders were unknown as replacements for the mortar and pestle. Sertanejos' archaic vocabulary (sometimes described more generously as "picturesque") reflected the region's long isolation, and ubiquitous poverty led to "the enslavement of the miserable by the few less ignorant individuals and those with some resources."[43] Many agricultural workers were effectively enslaved to their employers, to whom they owed insurmountable debts. Significantly, the one estate that impressed Penna and Neiva belonged to Englishmen who had constructed a twenty-league paved road to market their manicoba, a forage crop similar to cassava.

Penna and Neiva portray Brazil's backland culture as violent and barbaric, influenced by remnants of what they term African "savagery."[44] At one stage they dismissed some of their own "comrades" before entering particularly desolate terrain, fearing for their personal safety.[45] Yet elsewhere they echo Euclides da Cunha's admiration for the sertanejo's adaptation to his harsh climate, remarking on the population's courage, vigor, agility as cattle herders and instinctive ability to navigate unfamiliar *caatinga* terrain. At times the doctors articulate an understanding of racial characteristics that is more biological than cultural, such as when they describe the relatively healthy population of southern Goiás, with its higher literacy rate and greater material comforts, as a "more balanced mix of white and mestiços" than the poor populations of harsher interior regions.[46]

Nonetheless, Penna and Neiva offer a paternalistic progressive vision in which science and medicine could intervene to relieve the needless misery of the sertanejo poor. They argue that the fatalism characteristic of sertão farmers is a product of social and environmental circumstances and could be overcome by introducing the technological progress underway elsewhere, while sertanejos toil "at the margin of civilization."[47] With appropriate government services, the doctors believe, sertão inhabitants would strive to improve their circumstances. Noting the illiteracy rate, estimated at 80–95 percent in various locales, Penna and Neiva write, "We observed a lively desire on the part of parents in trying to educate their children, for it is common to see itinerant teachers who settle on *fazendas* for a time, charging 3$ monthly per student. . . . [Such examples] indicate once again the ardent desire of those people to escape from illiteracy, which the unconcerned public authorities do not even try to combat."[48] The tuition cost cited above amounted to a significant fraction of many adults' wages. Civil registries of births, deaths and marriages were very incomplete, and Penna and Neiva's report repeatedly condemns the lack of state administrative presence in rural Brazil. The Catholic Church was more in evidence than the state, thanks to peripatetic Dominican friars; many people whom the travelers encountered did not recognize Brazil as a political entity at all, considering Ceará, Piauí and the other northeastern states to be separate lands.

In order to make the sertão prosper, the doctors argue, the federal government should support empirical research into its social organization and natural resources. A stronger administrative presence would further national progress, providing education and essential infrastructure to Brazil's abandoned hinterland. Penna and Neiva's central message is that Brazil's backlanders could become positive contributors to a modern nation, but they need guidance and updated technology in order to advance: "The abandoned state in which the peoples of central [interior] Brazil remain exacerbates the natural spirit of routine behavior that governs them. . . . They are practically impervious to progress, since in areas where industrial artifacts [such as sewing machines] are sold at prices well within reach of a large number of laborers, these are rejected for a thousand and one reasons."[49] The physicians accuse southern Brazil of treating the northeast like a colony, levying taxes on livestock produced there but returning only one quarter of this revenue in expenditures on roads, railways, and other public services. This is, they assert, the opposite of how the republic's elite ought to interact with citizens who require their tutelage.

In making specific recommendations to IOCS, Penna and Neiva focused on three issues: deforestation, water supply, and disease. They asserted, without recourse to historical evidence, that the sertão's aridity had been exacerbated by the practice of setting fires to clear fields and improve pasture. The construction of rail lines had also depleted vegetation across a wide area. Penna and Neiva called for a survey of the remaining flora to serve as the basis for a development scheme that would take advantage of native botanical species: "Only with the help of scientific research can we know with assurance the economic potential of the northeast region and the means to develop it and exploit the natural resources which it happens to possess, placing man in a situation to dominate the environment through exact knowledge of all factors . . . which exercise a remote or direct influence on the development of a modern civilization, among populations that over more than three centuries have assimilated almost none of the great transformations at work throughout the universe."[50] The predominant agricultural crops of corn, beans, rice, tobacco, and sugar did not provide an adequate economic base for the population, and the physicians placed great hope in expanded cotton cultivation as a source of economic improvement, noting, "On the day when they decide to confront [the current state of misery] seriously, studying the species and varieties [of cotton] most suitable to the soil, this will provide a great transformation and bring abundance."[51]

Penna and Neiva were adamant that reservoir construction alone was insufficient to combat the suffering caused by droughts. They concerned themselves with many aspects of rural life, including crowded living conditions in mud houses without screens or glass on the windows, infrequent bathing, and the population's lack of shoes and clean water. Disease transmission through water was a recurrent preoccupation in their report, especially since it was popularly believed that illness could *not* be caught from contaminated water. In the absence of piped water, local watering holes were used for everything from bathing animals to drinking. The physicians warned that reservoirs constructed by the drought agency could become insect breeding grounds, transmitting diseases that did not otherwise thrive in the semiarid sertão.[52] They cautioned IOCS to keep new reservoirs free of vegetation on which mosquito larvae could grow.

Like Torres and Bomfim, Penna and Neiva believed that disease was rampant in the sertão due to the region's poverty, ecology, and political shortcomings rather than the population's inherent racial weakness. They found that quinine to combat malaria and common vaccines for humans and animals were virtually unknown. Popular remedies abounded, from cures based on religious beliefs and "superstition" to the use of plants whose "therapeutic action . . . is greatly exaggerated."[53] While Penna and Neiva acknowledged the efficacy of a few folk treatments derived from local botanicals, they were highly skeptical of the power attributed to popular healers who, they contended, could not themselves explain the source of their miraculous abilities. Some local medicines caused severe diarrhea or vomiting, further harming the sick. The physicians proposed that the federal government organize an itinerant medical service and pharmacy to bring basic modern therapeutics to the rural northeast. They recommended that a bacteriologist accompany the service, to study "obscure and poorly understood diseases present [in the sertão], which merit more adequate research."[54]

Malnutrition contributed to the high level of illness in the sertão. Many rural Nordestinos believed that fruits caused fevers and therefore did not

Small reservoir remaining in a riverbed during a drought, used by humans and animals. Photographed during the 1912 expedition led by Belisário Penna and Arthur Neiva. Source: Imagem IOC (AC-E) 2-15.10, Acervo da Casa de Oswaldo Cruz, Departamento de Arquivo e Documentação.

consume the wide variety available. Cattle roamed freely and were not milked, and few households kept poultry, fished, or planted gardens. The poor subsisted in a state of "painful misery" on whatever food they could find: honey, game animals, beans, manioc flour, rice when it was available, and coconut when in season. Salt and coffee were too expensive to procure due to transportation costs. Bread was essentially unknown, and tobacco smoking was a common means of assuaging hunger. Some families ate only one daily meal of manioc flour mixed with honey.

Fundamentally Penna and Neiva viewed sertanejos as a vulnerable population to whom Brazilian authorities had a moral obligation that must be met through the extension of public services and modern infrastructure into the sertão. This point was made eloquently in one anecdote:

One time, in a Bahian dwelling at considerable distance from any settlement, we had from the proprietor the precise definition of what exactly the sertanejo is, isolated from the world, without resources, without means of communication, telegraphs and [reliable] post, where news of what is happening on the planet is transmitted orally by the rare traveler who passes by. . . . In considering the material difficulty of conquering distances and populating those deserts, noticing that the day never came when a railroad finally passed his way, that although he is old he could not discern the slightest change for the better from when he was a child, and certain that his grand-children would die at a venerable age leaving things just as they found them, the man ended up concluding resignedly with a painful but truthful image: "this here is an open tomb."[55]

Such cultural and political lethargy is what the sanitarians believed modern technologies and medical science could ameliorate, accelerating social progress and redeeming the sertanejo race.

### Establishing a Rural Sanitation Service

Shortly after Penna and Neiva drafted their northeastern health survey, the president of Brazil's Academia Nacional de Medicina characterized the country as "an immense hospital," a nation in a precarious state.[56] Physician Miguel Pereira accused Brazil's politicians of neglecting their responsibilities to the poor, and he questioned the justice of recruiting soldiers who had never been offered elementary education or basic health care. Pereira "diagnosed illness as the nation's principal problem and elite disregard as the reason why little had been done to solve it."[57] Historians have emphasized the impact that this speech by a prominent physician had on public opinion, helping to launch a national public health movement. Penna was among those at the center of this movement, which gathered momentum through the early 1920s.

Penna's tour of the northeast inspired him to make a similar survey of Brazil's southernmost states at his own expense the following year. This experience permitted him to portray Brazil's rural health needs in national terms, more compelling to southern elites than an emphasis on only the distant northeastern sertão. In 1916, with private funding from supporters of his rural sanitation agenda, Penna established two health posts in the state of Rio de Janeiro. President Wenceslau Brás visited one of these after reading newspaper essays by Penna promoting a national rural health campaign, and in May 1918 authorized the creation of the Serviço de Profilaxia Rural (Rural Prophylaxis Service) for the federal district (surrounding Rio de Janeiro) under Penna's direction.

Later that year, health conditions in Rio were thrown into chaos by the Spanish flu; twelve thousand people died from the outbreak over three months, demonstrating the inadequacy of existing public health services even in the capital city. The epidemic struck as far north as Ceará, where the state secretary of internal commerce and justice petitioned the governor to adopt "a public hygiene regime satisfying the prescriptions of modern science, especially with regard to the interior," since illnesses constituted "terrible enemies of our prosperity."[58] President Brás accepted the recommendation of a National Academy of Medicine commission to make Penna's Rural Prophylaxis Service a national organization, overseen by the Ministry of Justice and Interior Commerce. The service was mandated to focus on endemic diseases as well as epidemic outbreaks. It continued the health post model established by Penna for the state of Rio de Janeiro.

One justification for expanding public health services throughout Brazil in this period was that Rio de Janeiro and São Paulo had benefited disproportionately from scientific and technological investment during the republic's early years. The publication of Euclides da Cunha's *Os Sertões* coincided with President Alves's program of physical improvements to Rio de Janeiro in the name of public health and increased foreign trade. Exporters in the capital pressured the government for reforms to make Rio a more appealing place to do business. The port was losing traffic to São Paulo and Buenos Aires because of disease outbreaks—especially yellow fever. Alves's administration took over sanitation of the federal district and cooperated with the city in implementing several renovation projects. These drew protest from some sectors of the population—particularly tenement dwellers who were forced to relocate, without government assistance, far from the city center when their homes were razed to make way for broad avenues (reminiscent of Georges-Eugène Haussmann's renovation of Paris).[59] Comteans also opposed the authority granted to government sanitation brigades, and they united with poor residents in riots protesting compulsory smallpox vaccination.[60] But physician Oswaldo Cruz, head of the federal government's Diretoria Geral de Saúde Pública (General Directorate for Public Health), successfully eradicated yellow fever from Rio within a few years by fumigating the mosquitoes that carried it. His campaign also lowered malaria incidence by reducing mosquito breeding areas. Cruz became a national hero when the Twelfth International Conference of Hygiene in Berlin awarded his institute of microbiological research a gold medal in 1907 for its contributions to hygiene science. Following Cruz's accomplishments, governing elites embraced modern medicine, rooted in germ theory and the emerging science of parasitology, as Brazil's best means to escape a long-held fatalism about their tropical nation's prospects.[61] Brazilian intellectuals' conviction that scientifically based administration was the key to national progress, along with their growing awareness of the disparity between the backland sertão and coastal *litoral*, led to greater interest in funding empirical investigations and development of the country's interior.[62]

Another reason the federal government expanded rural health services during this period was that doing so enabled some assertion of authority over state affairs without disregarding the constitutional limits on federal power. Several historians have argued that public health services became a wedge for extending federal agencies into Brazil's interior in hopes of reducing rural oligarchs' stranglehold on local politics. As Gilberto Hochman notes, illness—particularly epidemics—spread due to the increasing territorial interdependence that accompanied internal market expansion. Such shared responsibility for disease seemed to justify a shared public response in the form of federal public health campaigns.[63] Most states that cooperated with the Rural Prophylaxis Service to establish health posts were unable to cover their stipulated financial obligations and ended up in debt to the federal government, which reduced the state's administrative autonomy. São Paulo was the only state to run a successful health and sanitation program without inviting federal intervention. The coffee-rich state's urban elites argued that a decline in immigration during World War I made their economy dependent on rural laborers, and better hygiene would increase the productivity of those workers. Despite opposition from conservative paulista planters, Arthur Neiva was hired to establish a state sanitary service at the end of 1916.

## Brazil's Liga Pro-Saneamento and the National Department of Public Health

To promote greater federal involvement in public health, Penna and other prominent sanitarians formed the Liga Pro-Saneamento (Pro-Sanitation League) in February 1918 on the first anniversary of Oswaldo Cruz's death. During the league's two years in existence, members included lawyers, doctors, engineers, military officers, and politicians—President Brás among them—who distributed pamphlets about hygiene, offered lectures in schools and public parks, published newspaper articles about the need for improved sanitation throughout Brazil, and provided health services to rural workers. League members emphasized hookworm, malaria, and Chagas disease (*Trypanosomiase americana*) as the country's most debilitating illnesses, outweighing smallpox and yellow fever epidemics in their damaging effect, especially in the countryside. They had close links to the Sociedade Nacional de Agricultura (National Agricultural Society) and advocated improving the health of Brazil's rural workers as an alternative to subsidized European immigration, which had ceased to be an effective strategy for labor recruitment.

The Liga Pro-Saneamento published a journal of limited circulation, *Saúde* (Health), for one year starting in August 1918. The journal's editorial board included Penna (a prolific contributor), Astrogildo Machado of the Instituto Oswaldo Cruz (leader of a 1912 sanitary expedition along the São Francisco River valley), several federal health inspectors, and anthropologist Edgard Roquette-Pinto. Contributors aimed to define a new national identity that drew on Brazil's heritage but incorporated modern knowledge and technologies; like other nationalists during the First Republic, they embraced the positivist notion that social progress required rational, scientifically based management. Following Alberto Torres, league members believed that industrialization was inappropriate in Brazil; it made sense only for overpopulated countries where people could no longer subsist from the land.

To raise funds for the league and promote its agenda, in 1918 Penna compiled his newspaper essays on the health problems of Brazil's interior in a book titled *Saneamento do Brasil*, with the subtitle *Sanear o Brasil é Povoal-o; é enriquecel-o; é moralisal-o* (To sanitize Brazil is to populate, enrich, and moralize it). The book attacks the republic's economic policies for favoring southern elites and industries at the expense of the rest of the nation and proposes an agrarian modernization strategy similar to that promoted by Torres. Throughout the volume Penna employs the term sertão expansively to refer to all of Brazil's insalubrious hinterland. Such a broad definition dates back to the colonial era, when Portuguese settlers used sertão to describe "vast unknown realms, remote and scarcely populated."[64] Scholars emphasize the frequent contrast between sertão and *litoral* (roughly, "interior" and "coast") in discussions about the problem of national consolidation. As linguist Jorge Amado notes, the term litoral encompassed "a known area, limited, colonized or in the process of being so, populated by other races (native Indians or blacks) but dominated by whites, a region of Christianity, culture and civilization;" sertão, on the other hand, delineated "unknown areas, inaccessible, isolated, dangerous, dominated by brute nature, and inhabited by barbarians, heretics, infidels, where the benefits of religion, civilization and culture have not arrived."[65]

Republican readers were most familiar with the term sertão as a designation for the northeastern hinterland. Penna's expansive use of the term
to encompass the entire landmass beyond Brazil's coastal urban centers was
politically astute. Asserting from his own experience that serious health
problems affected the entire countryside bolstered Penna's argument for the
need to finance and administer a federal public health service. Brazil's
southern elites could ignore the problems of the sertão if that encompassed
only the northeastern hinterland, a region that they viewed as culturally
backward and racially dubious. But no Brazilian with a stake in his country's modernizing ambitions could afford to overlook health issues crippling
much of the nation's rural population. As Hochman observes, the term sertão
as it came to be employed by sanitarians during the latter 1910s was "more
a medical, social and political category than a geographic one. Its spatial
locale depended on the existence of the binomial 'illness-abandonment.'"[66]
Even the states of Rio de Janeiro and São Paulo had neglected sertões just
outside the borders of their capitals.

*Saneamento do Brasil* introduced themes that Penna reiterated in various speeches and essays during the following decade. He defined hygiene
as a social and political as well as medical issue and emphasized the vicious
cycle linking illness and poverty. He argued that the apparent indolence of
Brazil's rural poor was the result of disease, not innate racial handicaps or
tropical malaise. He noted that Brazil's high incidence of tuberculosis
stemmed from the physical weakness caused by malnutrition combined
with other diseases. Penna leveled his harshest criticism at the republic's
governing elite for their bias toward the industrializing south, portraying
subsidies for urban industry as an unjustifiable burden on the national economy that served wealthy southerners' interests while neglecting the health
and well-being of most citizens.

Penna's proposals for national economic development echoed Torres's
recommendation of increasing agricultural production to reduce Brazil's dependence on foreign investment, which industrialization relied upon. In
Penna's colorful metaphor, the Brazilian republic was a house with a well-
furnished parlor (São Paulo) but otherwise in a precarious state, having invested all of its resources in one area in order to impress visitors.[67] To
correct these inequities, Penna demanded renewed investment in rural
areas, where many workers had been neglected since the abolition of slavery in 1888. He repeatedly compared the republic unfavorably with the
monarchy that preceded it (from independence in 1821 to the end of the
empire in 1889), and he wrote nostalgically about a pastoral past in which

free and slave workers lived happily off the land, cared for by paternalistic landlords.

Penna essentially proposed scientific paternalism as a replacement for the paternalism of the prior slave system, when workers had been valuable to their owners and consequently (he imagined) well cared for. He envisioned modern Brazil as a nation guided by scientifically trained men like himself in a development process suited to Brazil's particular climate and population, independent of foreign capital or expertise. Hygiene science, combined with civil engineering, would rationally incorporate the sertão—expansively defined—and its inhabitants into modern economic and social life. In the process, the federal government would extend its reach into rural areas, which Penna viewed as a necessary means of weakening the hold of landowning *coronéis* on their client populations. At his most ambitious Penna hoped that expanded public health services would spawn a political revolution in the retrograde interior.

The Liga Pro-Saneamento's plan for rural sanitation appears as the final essay in *Saneamento do Brasil*. The proposal was a radical departure from the republic's 1891 constitution, which granted the federal government authority to interfere in states' business only in limited ways (such as to oversee port sanitation); it called for the establishment of a federal health department with authority throughout Brazil, backed by the federal judiciary to circumvent the capriciousness of state courts and administrations. A sanitary code would stipulate the responsibilities of agricultural and industrial employers for workers' health, along with other matters.

In concluding his book, Penna called for a series of reforms. The "momentous sanitary problem," he wrote, "is not just medical and hygienic, but also social, political, and economic." Along with treating the sick and improving their houses (by installing cesspools and wells), the state needed to supply communication technologies, elementary schools, and agricultural extension services as well as reduce prices on essential goods. For Brazil's rural population to reach its full economic and civic potential,

> The most active, incessant propaganda must be conducted to inculcate in the spirit of all our patricians, in particular our public men, men of letters, journalists—intellectuals, in short: that alcoholism, *Trypanosomiase americana*, malaria, and hookworm to a great degree, and other endemic illnesses to a lesser degree, are the cause of our backwardness and of the shameful rearguard which we continue to

occupy relative to other peoples. . . . When we liberate our people from these scourges, they will not envy any other people for their robustness, capacity to work, and intelligence. . . . They will shake the ignominious yoke of the satraps and factions; become conscious of their rights and dues; acquire love of country; and make themselves to be governed conveniently, by capable men, of their own free choosing.[68]

Penna expected an expansive public health program to inspire profound social transformation. People relieved of bodily suffering, he prophesied, would have the energy and self-respect to eliminate other pernicious aspects of national life that limited their prosperity and the country's progress. Elsewhere in the book he warned that Brazil's "sons of a fictitious industrial progress" were sorely mistaken in assuming that the country was securely theirs: "They are unaware of or do not remember the biological phenomenon of hibernation or latency, observed at each step in nature and history, revealed in '89 in France, recently in Mexico, and unexpectedly, a few days ago, in Russia."[69]

Penna had numerous supporters, particularly among physicians, and many published their own observations and conclusions about the need for rural sanitation. Pernambucan physician Octavio de Freitas cited Penna and Neiva's northeastern health survey in a 1918 lecture on the problem of "anemia" among rural workers, in which he vindicated that population as ill rather than lazy. Freitas's own surveys over the previous decade had confirmed the prevalence of malaria and hookworm in Pernambuco's interior, and he encouraged the state to pursue a program of rural sanitation rather than focusing public health resources on the less plagued capital.[70] In a speech inaugurating Paraíba's Serviço de Saneamento Rural in 1921, physician Acácio Pires traced the history of Brazil's rural sanitation movement from Miguel Pereira's hospital metaphor through Penna and Neiva's report, the health posts established by the Rockefeller Foundation's IHB, and Penna's rural sanitation service. He concluded with a statement emphasizing the state's role in public health: "Saneamento is not just a work of medicine and hygiene, but is above all a political project, relying on the government to unite every effort in order to achieve it adequately."[71]

Under pressure from the Liga Pro-Saneamento and its supporters, newly elected president Epitácio Pessoa—the first Nordestino to occupy that office—reorganized federal public health services at the end of 1919. Pessoa established the Departamento Nacional de Saúde Pública (DNSP; National Department of Public Health), directed by physician Carlos Chagas, who had discovered the parasite responsible for the disease that bears his name in 1909.[72] The Liga Pro-Saneamento ceased to exist after the DNSP was created, since it had achieved its main goal, but several former members became key figures within the new federal health administration. Chagas nominated Penna to head the new Diretoria de Saneamento e Prophylaxia Rural, essentially continuing his previous position with the Rural Prophylaxis Service. Along with establishing health posts, the rural hygiene directorate employed engineers to inspect building construction, water supply, and sewage disposal for compliance with new national health and safety regulations (though these were not widely enforced).[73]

The creation of a federal public health department awarded sanitarians greater authority within the Ministry of Justice and Interior Commerce, which previously had limited responsibility for protecting public health during epidemics. Nevertheless, because of constitutional restrictions on federal interference in state affairs, the new department was required to reach an agreement with each state about cooperative financial and administrative responsibilities for public health efforts. The DNSP provided half of the funds for rural health posts if its Diretoria de Saneamento e Prophylaxia Rural had administrative control over them; if states remained in charge of the posts, the federal contribution was one-third.[74] Penna strongly objected to the requirement that states contribute at least half of the costs for rural sanitation work. He feared that relying on state contributions would exacerbate economic and political disparities between states and further diminish national unity, since wealthier and better-administered states would improve the health and productivity of their populations while others lagged behind.

Penna's proposal that the DNSP's Diretoria de Saneamento e Prophylaxia Rural be financed through a "health tax" on *cachaça* was legislated in January 1920. This alcoholic beverage, fermented from sugarcane, was the ubiquitous solace of the poor, and Penna viewed it as an added source of unproductivity and domestic instability in rural Brazil: "In any shack or hut in the interior, lacking everything, even food, there is a bottle of the drink, with which the wretched resident suppresses hunger and sadness."[75] Penna recommended that the government distribute half of the revenue from the alcohol tax equitably among all states based on their land areas and population. The money could be used to establish rural health posts, ensuring that improvements would be witnessed throughout the country; states that wanted to expand these services could reach an agreement with the DNSP to split the cost of additional posts if funds were available. The remainder of the alcohol tax should, Penna proposed, be used for sanitary education campaigns, new agrarian settlement in strategically situated areas, and addressing emergencies such as epidemics or food shortages. In this way Penna hoped to buffer his national sanitation program against regional political differences, promoting social betterment throughout rural Brazil.[76] This proposal to change how rural health funding would be distributed among states was not adopted, however.

Hygiene inspectors in the northeast enthusiastically embraced the federal government's offer to subsidize rural health services. José Moreira da Rocha, Ceará's director of hygiene, wrote hopefully in 1919 that his state's rural population could be rescued by modern sanitation and medical care with the addition of resources beyond the state's coffers. He echoed Rio de Janeiro's and São Paulo's sanitarians in asserting that the sertanejo population was burdened by disease rather than debilitating racial endowment or immutable cultural deficiencies:

A series of intelligent decrees has established regulations for Union intervention in rural sanitation of the states. Hookworm, plague, trachoma and malaria are in our backlands, hills, and beaches, crying to the state government to take advantage of those salutary federal resolutions and their arrangement with the Union or the Rockefeller Foundation in order to begin now a perfect and rational preventative course in place of the mere assistance or parsimonious distribution of medicines that we have achieved until now. *That the health of our rural worker is the basis for the efficiency of his work, of our wealth and of our progress, it is unnecessary to repeat. Our sertanejo is more an invalid than a layabout, as they say; someone incapacitated by disease and by ignorance, but not by atavistic or regional inferiority—this has been demonstrated.* It is thus rural sanitization that comprises an essential part of instruction, sanitary education the most relevant problem to resolve, the most elevated campaign to undertake, the most noteworthy deed to accomplish in our environment, from the point of view of hygiene and general progress.[77]

In the following year, with the assistance of a federal sanitary commission, Ceará installed a bacteriology lab for its hygiene directorate and established a Pasteur Institute for the production of serum to combat rabies—the latter funded largely by private donations, as was the case for similar institutes worldwide. The state hygiene director traveled to Rio for four months to study bacteriology, "the science which is today the soul of hygiene."[78] Ceará's sanitarians had entered the world of modern public health, circumventing state inaction through support from private and federal sources.

Historian Luiz Antonio de Castro Santos has argued that Brazil's public health bureaucracy of the 1920s, despite its achievements in rural areas, actually represented a conservative modernization strategy.[79] The federal government bolstered many state political machines in return for their compliance with its sanitary reforms. Public health administration was the federal government's route to consolidating its power in previously autonomous regions, but the smoothest way to do this was by cooperating in large degree with state governing parties. This helps to explain why the advance of sanitation services did not spawn a broader transformation of rural society, as Penna had hoped it would; his estimation of the leverage that improved health would provide to effect broader social reform was overly optimistic.

## The Rockefeller Foundation's Rural Health Campaign in Brazil

The Rockefeller Foundation's International Health Board was also involved in funding rural health posts throughout Brazil by 1919, and the nation received more than half of the aid provided by the IHB in Latin America during the organization's early years.[80] The IHB surveyed sanitary conditions in São Paulo in 1915 and began a rural health campaign in September 1916 that was led by physician Lewis W. Hackett of Harvard University. (Hacket would remain with the IHB for thirty-five years.)

Until 1919, IHB personnel in Brazil had focused exclusively on southern and central states, beginning with Rio de Janeiro. IHB staff took a dismal view of northern Brazilians, perceiving the south's white immigrant population (particularly around São Paulo) to be more "vigorous" and receptive to development initiatives. The IHB's priorities and ideology closely mirrored those of republican Brazil's conservative modernizers, who supported "whitening" immigration policies as the quickest route to economic expansion. This stance contrasted notably with Penna's analyses of Brazil's health crisis, which focused on poverty and inequality as the root causes of disease. Penna's relative sympathy for the plight of rural workers is absent from the reports and correspondence of high-level IHB staff who oversaw the Rockefeller Foundation's Brazilian health programs during these years.

As in all places where it worked, the IHB's main goal in Brazil was to demonstrate efficient, effective methods of sanitary administration that could serve as a model for public health programs independent of foreign aid. The health posts established by the IHB in its first years in Brazil focused almost exclusively on hookworm eradication, the board's signature "demonstration" project that had achieved notable success in the U.S. South and elsewhere.[81] Fostering a scientific approach to public health administration was the IHB's primary goal, of greater importance than curing specific diseases or patients. "The actual benefit conferred upon the infected population is not an aim but a side-product of the Board's activities," Hackett explained in 1919; "the main purpose of this work [in Brazil] is education and not practical or utilitarian."[82] IHB director Wickliffe Rose reinforced this agenda regularly, telling Hackett, "When federal and state authorities have learned the lesson of efficient administration, the end will have been achieved."[83] Brazilian IHB personnel sometimes transferred to state or federal health administrations; as Hackett noted following the resignation of one staff member to direct a federal sanitary commission, "Their training and altered viewpoint is not the least important by-product of our work in Brazil."[84]

IHB staff carefully selected regions where their demonstration of the advantages to be gained from public health initiatives would be most persuasive. In Brazil, board staff quickly concluded that better public health administration was badly needed throughout the country, especially in rural areas. Hackett told the New York headquarters to advise incoming staff,

> A Field Director going into a county in the State of Rio de Janeiro . . .
> will find no health department, no practicing physicians, no roads, no
> hospitals, no health laws, no public opinion, no social feeling, no
> statistics, no food, milk or water inspection, no medical school
> inspection and no compulsory school attendance. The illiteracy will
> run to about 85% of the population. . . . He will find the people
> docile and results easy to demonstrate because the conditions are so
> bad. I am merely trying to get at the idea that the work will not be
> complicated here and that it will require a thorough knowledge of
> the fundamentals, particularly as regards hookworm and latrines,
> malaria and superficial drainage, smallpox and venereal diseases.[85]

Given the widespread need, the IHB's central administration instructed Hackett to establish health posts in areas with strong state and municipal governments, where the effectiveness of their model could be most impressively demonstrated. São Paulo offered the most suitable government structure for fostering a system of municipal (county) health units, which was the ultimate aim of the IHB's hookworm campaign; it was followed by Rio de Janeiro and other southern states.[86] The municipal health units were expected to undertake home, water, and school inspections; run a dispensary and bacteriological laboratory; maintain epidemiological and vital statistics; and promote general sanitary education. As historian Lima Rodrigues de Farias argues, "The Rockefeller Mission's areas of activity reflected and reinforced [Brazil's] regional differences" by concentrating resources in the rural areas of wealthier and better organized states.[87] This was the opposite of what Penna recommended as the first director of the Diretoria de Saneamento e Prophylaxia Rural.

The IHB was reluctant to cooperate with northeastern states in part because it feared that they would not uphold their part of shared financing agreements for rural health programs, but federal willingness to subsidize states' obligations to the board increased the appeal of launching health posts in the northeast. In early 1919 the new Serviço de Profilaxia Rural offered to absorb a quarter of the cost of IHB projects, with the IHB funding an equal portion. This left half to be covered by states, though in many cases they did not fully meet their obligations. With the creation of the DNSP at the end of 1919, the federal government assumed half of the cost of rural sanitation work for poorer states, making IHB projects feasible in the northeast (with the board continuing its policy of covering one quarter of the expenses).[88] In order to obtain federal aid for rural public health work, states had to adopt a sanitary code modeled on the one developed by DNSP director Carlos Chagas. Since the DNSP required health posts that it supported to address two endemic diseases, the IHB agreed to add antimalaria services (quinine distribution and drainage efforts) to its hookworm treatments.[89]

Bahia and Pernambuco were the first northeastern states to take advantage of the collaborative arrangement involving the DNSP and IHB in 1920; in the following year it was adopted by Alagoas, Ceará, and Paraíba. Bahia was surveyed at IHB expense by Brazilian doctors, since a yellow fever outbreak along the coast in 1920 made it temporarily unsafe for nonimmune American personnel. At that point the IHB employed five American doctors to oversee the health post program, thirty-two Brazilian doctors who served as field directors (and would eventually take the Americans' place), and 350 Brazilian sanitation assistants. The medical surveyors found 96 percent of the population to be infected with worms.[90] The IHB also sponsored a Brazilian physician to survey Maranhão (the state bordering the Nordeste and Amazon regions), and he found that over 90 percent of the population suf-

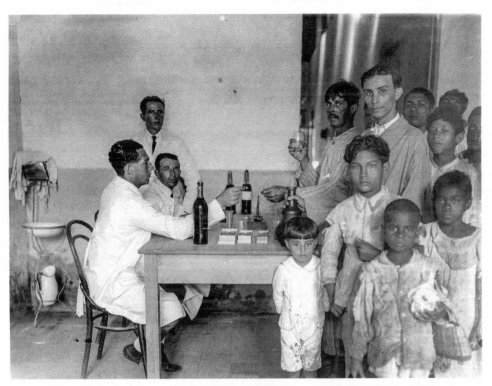

Doctors and patients in the pharmacy of a federal rural health post in Paraíba, ca. 1923. Source: Imagem BP (F-VPP) 21-6, Acervo da Casa de Oswaldo Cruz, Departamento de Arquivo e Documentação.

fered from hookworm. In response, both the IHB and the federal government opened laboratories in Maranhão for bacteriological analysis and hookworm treatment.[91] Maranhão's government also established two health posts to treat parasites, reduce soil pollution, dispense quinine, vaccinate against smallpox, and conduct hygiene education. The IHB began a survey of Pernambuco in March 1920 under the direction of physician Fred Soper; he found that 97 percent of the population surveyed (in the eastern half of the state) had parasites.[92] Pernambuco had established one rural hygiene post in 1918, after conducting its own hookworm study, which the IHB now began directing along with new ones.

Despite the substantial disease burden revealed by these northeastern health surveys, IHB director Rose expressed strong prejudices against Brazil's Northeasterners in his 1920 report on the board's Brazilian program. Rose viewed São Paulo as a dividing line within the country marking contrasts "as sharp as those between Mexico and the United States." He believed that mixed-race Nordestinos would never attain the same level of social development as their compatriots in the South, where European blood had come to prevail over African and indigenous influences. Rose was an advocate of the "whitening" immigration policy promoted by paulista elites. He shared their racist skepticism about the Nordestino population, which he characterized as "composed of shiftless blacks and parasitic whites of Portuguese origin, and a large percentage of their hybrid progeny." Ceará was the exception to this bleak picture, having been founded by hardy adventurers who "intermarried with a particularly virile native Indian tribe and developed a sturdy native population. In the face of extremely hard conditions of periodic drought and famine, these people have remained in their state and are today an energetic self-reliant race."[93] In Rose's view, inhabitants of the demanding drought zone had benefited from a process of natural selection. With little humid coastal territory suitable for sugar cultivation, Ceará's inhabitants had a smaller proportion of African ancestry than the populations of northeastern states (such as Pernambuco) where the colonial plantation economy had been centered. The racial intermixture of whites with native índios was evidently more palatable to Rose than *mestiçagem* involving blacks, a view expressed by many Brazilian writers at the turn of the twentieth century—most notably novelist José de Alencar.

Praising the South Brazilian population in contrast to their Nordestino compatriots, Rose described the

> adventurous, self-reliant Portuguese, who from the beginning
> crossed with the native Indians, developed a sturdy Brazilian stock,
> established themselves on the narrow coastal margin at Santos [São
> Paulo's port city], and proceeded at an early date to explore and
> conquer the interior. This population has been re-enforced through a
> tide of immigration from Europe which continues to bring these
> southern states hardy types of colonists—Italians, Germans, Austrians,
> and Poles. Japanese also are now coming in considerable
> numbers. These immigrants take root in the soil, and tend in the
> second generation to become a sturdy, white, Brazilian stock. . . .
> These southern states, having the advantage of a cooler and more
> variable climate and of a vastly more virile population, have in their
> keeping the future of Brazil. *It is the self-reliant white man who is
> pushing back the frontier and laying the foundations of a more progres-*

*sive civilization. The State of São Paulo is the center and soul of this movement . . . . The hope of the North lies in the South's leadership, and in new blood from these States and from Europe.*[94]

Unlike Penna, who argued that the low productivity of many rural Brazilians—especially in the Northeast—was caused by disease pathogens, Rose articulated a racial explanation for regional differences and believed that "new blood" was needed to invigorate national development. In this view, Brazil's rural backwardness was not simply cultural but fundamentally hereditary. Modern science and technology, rationally administered, were necessary but not sufficient to precipitate rural progress.

Rose's analysis of Brazil's modernizing potential contrasts sharply with Penna's. Penna, along with other Brazilian nationalists in this period, sought to end subsidized immigration of Europeans and focus national resources on the agricultural workers already struggling in Brazil. He believed that his country's rural poor were burdened by illness and illiteracy more than racial inferiority. His tours of the country's interior alerted him to the plight of individuals whom he saw as abandoned by public institutions that had a civic responsibility to keep step with medical and technological advancements. Rose, on the other hand, in assessing Brazil's prospects for rapid economic development, saw São Paulo as the region most able to increase the country's industrial output due to its population's European biological and cultural influences. As the IHB's director, Rose viewed Latin America as a set of national economic units and U.S. trading partners. He aimed to increase each country's overall productivity and consumer base through improvements to public health in the regions that could most readily adopt them.

Rose and Penna thus offered contrasting recommendations for how public health should be organized in Brazil. Rose believed, based on São Paulo's experience, that health administration should be brought under state and local control. He assumed that this would lead to greater efficiency in aiding ill populations. Penna favored more centralized, federal control over public health on the grounds that the neediest populations lived in states where local authorities neglected the public good. Historian Nísia Trinidade Lima interprets Rose's enthusiasm for local governance as the misplaced imposition of a U.S. model, where rural health had become the responsibility of county and town administrators under Rockefeller Foundation tutelage.[95] In areas like Brazil's northeastern sertão, without democratic local governments, strengthening municipal and state authority often meant reinforcing oligarchic tyranny and abetting disenfranchisement of the landless poor.

## The Sertão and Scientific Modernization

In the 1910s, Brazil's sanitarians helped to shift national discourse about rural workers away from speculations concerning permanent racial and climatic handicaps and toward a discussion of how science could modernize the country's expansive interior. Beginning with the report he coauthored with Arthur Neiva, Belisário Penna consistently laid blame for the apparent backwardness of rural laborers on the negligence of national and regional leaders. Many other sanitarians shared Penna's faith in the social progress that public health measures could engender. Their goals were supported by opponents of republican oligarchs, patriotic nationalists who felt that Brazil's leaders had failed in their obligation to invest in modern improvements—particularly outside of urban centers. But despite sanitarians' success in reorganizing federal health administration, they struggled to persuade powerful southern politicians that national investment in the northeastern sertão and other hinterland regions was worthwhile. Leaders of the Rockefeller Foundation's International Health Board remained similarly unconvinced that Brazil could promote economic modernization through improving the health and security of mixed-race sertanejos rather than by devoting resources to descendants of European immigrants—predominantly in São Paulo—as branqueamento ideology prescribed.

Penna's approach to sertão modernization was integrationalist; he proposed federal paternalism as a solution to sertanejos' multiple woes in return for their increased incorporation into the national economy. This may not have been many sertanejo farmers' ideal; regional popular poetry of the period often expresses the simple desire to live independently on a small farm, with little concern for personal wealth or national progress. But Penna can be credited for his insistence that all sertanejos, as legitimate citizens of the modern republic, should be provided with basic health and economic infrastructure and freed from dependence on rural estate owners.

· · · · · ·

Chapter 3 will implicitly contrast nationalist Brazilian sanitarians' concern for sertanejos' welfare with civil engineers' proposals for dam and irrigation works to alleviate the scourge of drought during the 1910s and 1920s. The ideological biases that influenced engineers' sertão development strategies are not immediately evident in their technical discussions of reservoirs and roads as solutions to the drought problem, but such biases—and alternatives to them—are clearly articulated in the contrasting writings of Wickliffe Rose and Belisário Penna during the same period. The drought agency's plans, modeled on similar projects elsewhere in the world, rarely questioned or confronted existing landholding and labor patterns in the Nordeste. As a result, the civil engineers' regional development efforts had the unintended effect of further concentrating land and water resources in the hands of relatively few landowners, to the detriment of the sertão's marginal poor.

Nonetheless, in the process of managing public works construction as a form of emergency aid to sertanejos, a number of the federal agency's technical staff came to understand the region's drought problem as comprising much more than climate instability. They increasingly argued that reducing the suffering caused by drought would have to involve political as well as technological reform. This growing realization—analogous in many ways to Penna's insights in the public health arena—is analyzed in chapter 3. Just as Penna came to see disease as a reflection of underlying social ills through his interaction with the rural poor, the managers of worksites during drought years came to see sertanejos' vulnerability to climate fluctuation as a product of social inequities. Their instinctive assessment, stemming from jarring interactions with starving families, is consistent with the theoretical framing of natural disasters put forth decades later by Amartya Sen and others: that the roots of such crises are fundamentally social and political.

**Engineering the Drought Zone**

The Birth of IFOCS, 1909–1930

· · · · · · · · · · · · · · · · · · · · · · · · · · · · · · · · · · · · · · · · · · ·

In February 1920, engineer Alarico Araújo sent the following urgent telegram to Brazil's drought works inspector in Rio de Janeiro. Araújo was overseeing construction of the Santo Antônio Reservoir near the town of Russas, Ceará. His manual labor force consisted of starving men who sought food rations for their families in the midst of a searing drought. "Due to the scarcity of potable water, the situation here is becoming truly desperate [*afflictiva*]," Araújo wrote. "Since October we have been transporting water from small watering holes along the Rio Palhano, 8 km from the reservoir site. At the beginning of this week, two of those *cacimbas* ran dry. Since then, the workers' thirst has been unquenchable. I am hastily arranging transportation of water from the Lagoa do Peixe, 12–14 km from here. This solution will result in additional expenses, and I would recommend suspending this construction effort." Araújo's brief missive to his superior enumerated the dam site's desperate circumstances and measures taken to alleviate them so far. They had tried drilling new wells after the initial one dried up, but subsurface rock made this too difficult with the equipment available. Araújo was already managing several hundred workers, yet bands of *retirantes* (drought migrants) continued to arrive daily, pleading for some means of obtaining food. He had already turned many families down. In February Araújo was permitted to open a new road construction project, to accommodate some additional indigent drought victims.[1]

Reports such as this one from the front lines of drought works in the semiarid sertão illuminate several significant features of those projects during Brazil's First Republic. Most obviously, the civil engineers who managed them oversaw projects that also served as humanitarian relief operations. The majority of their workers were farm laborers with little experience constructing dams, roads, or irrigation works. These men (it was agency policy to employ male heads of households) were near starvation, and it was this that had compelled them to walk to the dam site, families in tow, hoping to exchange physical labor for subsistence food rations. Engineers responsible for these operations, such as Araújo, were often aghast at the desperation of the migrants enrolled in their projects. Numerous communications from construction managers to their superiors in Fortaleza or Rio de Janeiro requested permission to raise the daily wages (*diárias*) provided to workers and emphasized that the minimal food that could be obtained using the established diárias put retirante families at grave risk of illness and gradual starvation. In many cases, engineers' appeals for permission to pay their construction workers higher wages, or to aid their destitute households in other ways, were bluntly refused by drought agency administrators far from the dam sites. Despite its de facto role as a relief agency, the drought inspectorate's official function was to oversee public works construction in the sertão on the cheap.

Archived communications between drought agency engineers and their superiors suggest that almost no one in the Inspetaria de Obras Contra as Secas (IOCS; Inspectorate for Works to Combat Droughts) contested the basic premise of its operations—namely, that it would expand hydraulic and transportation infrastructure in the sertão by taking advantage of landless and smallholding agricultural workers' desperation during droughts, when they were temporarily unable to sustain their families without federal assistance. Even the managing engineers most sympathetic to the plight of their workforce, and most willing to advocate on their behalf, confined appeals to their immediate superiors to requests for wage increases; they did not directly question the priorities and professional hierarchy of the drought inspectorate, which established construction sites where hospitals might have been more beneficial.

Yet historical records of sertão drought works complicate the standard historiography of an uncaring federal "drought industry." Particularly in crisis years such as 1915 and 1919, the inspectorate's field engineers found themselves in bizarre and uncomfortable positions in relation to the sertanejo poor. The unflinching stance of IOCS was that its aid could only be provided in return for physical labor; in the words of Aarão Reis, agency director from 1913 through 1919, "charity degrades and corrupts, while work ennobles and fortifies character."[2] Thus, male heads of migrant families were offered a minimal daily wage—often in the form of credit to obtain food from middlemen (the infamous *fornecedores*)—to keep their families from starvation. The engineers who managed these worksites often wrote in frustration to their superiors, noting that the diárias they offered could not compete with any other work options in the sertão; once rains returned, or if any other industry (such as carnauba palm wax cultivation and extraction) was hiring, their workforce diminished precipitously. Labor for the

Drought agency reservoir under construction, 1912. Note that workers are carting rocks and dirt in wheelbarrows. Source: Imagem IOC (AC-E) 2-16.2, Acervo da Casa de Oswaldo Cruz, Departamento de Arquivo e Documentação.

federal drought agency was designed as a last stand against starvation and a dirt-cheap way for the government to improve regional hydraulic and transportation infrastructure.

This chapter explores IOCS's framing of drought crises as the result of climatic phenomena rather than socioeconomic dynamics during the inspectorate's first two decades in operation. Although construction site managers confronted the sertanejo poor directly and were acutely aware of their hardships, the agency's directors, based in Rio de Janeiro, understood the sertão primarily through maps, numerical data, and brief visits to dam sites. They concerned themselves with the region's geography, hydrology, and transportation networks and knew little about the living conditions or quotidian struggles of rural workers. Popular portrayals of poor sertanejos, most notably in Euclides da Cunha's epic *Os Sertões*, depicted them as a race apart, prone to religious fanaticism, superstition, and barbaric violence. For elite members of a profession that prided itself on rationality in all things, such perceptions must have made lowly sertanejos seem very foreign. Maintaining a firm belief in the transformative power of technology regardless of the larger social landscape in which it operated aligned with Brazilian engineers' positivist training.

IOCS administrators' technocratic response to the drought problem was also colored by their desire to serve as advisers to the state, thereby elevating their profession's stature and employment opportunities. This goal was best attained by offering solutions palatable to the politicians who governed the federal inspectorate's funding, and it would have been jeopardized by meddling in regional politics on behalf of rural workers. During the First Republic, local oligarchs dominated political life, particularly in the northeast. Candidacy for legislative positions was determined by family-based electoral machines, and leaders of state parties extracted federal resources in return for delivering votes during presidential elections. Northeastern landowners were resistant to changes that could diminish their control over natural and human resources, and southern politicians resented the siphoning of federal funds to a region they perceived as backward. Within this context it was expedient for the drought inspectorate's managing engineers to recommend the improvements to transportation, water management, and agricultural production desired by their northeastern patrons in the national legislature. Rather than acknowledge that their works would affect sertanejo social classes differently, depending on access to and control over land and water, proponents of expanded drought works claimed that their projects addressed broad "regional interests." In fact, many of the agency's constructions helped to increase the wealth and influence of rural *coronéis*.

## Brazilian Engineers as Development Advisers

Engineers were involved in discussions of the drought problem from the time that it became identified as an important national issue in the 1870s onward, but their proposals were not always welcomed by Nordestino politicians. During the Great Drought of 1877–79, the imperial court appointed engineers to study the sertão, and northeastern elites opposed the resulting plan to create irrigated farmland along riverbanks by constructing large reservoirs surrounded by colonies of smallholders. Instead landowners lobbied the government to subsidize smaller reservoirs on private property, arguing that these would stabilize the sertanejo population across a larger area. Not incidentally, smaller reservoirs eliminated the need for land expropriation by the federal government, contributed to increased property values, and retained rural workers on existing estates.

The establishment of IOCS in 1909 within the newly created Ministério de Viação e Obras Públicas (Ministry of Transportation and Public Works) gave engineers a stable institutional setting from which to address the drought problem. The new agency and its supervising ministry promised increased funding for public works and expanded employment for civil engineers. Yet many northeastern power brokers grew impatient with the inspectorate's surveys during its early years; they were eager to take advantage of drought aid to fund infrastructural improvements for their states and accused the federal government of employing scientific experts to delay more substantial investment in the sertão. The first generation of IOCS engineers described their efforts as involving constant struggle with negligent or corrupt politicians.[3]

During the nineteenth century, Brazilian engineers had found employment building railroads and other infrastructure to aid the south's thriving export economy. Their profession had long been tied to the state via the military; the first civil engineering school, the Escola Central in Rio de Janeiro, was originally the Escola Militar, and its renaming in 1858 signified the expansion of civil engineering beyond military functions. In 1862 the Ministry of Agriculture, Commerce, and Public Works created an engineering corps that comprised the first major public employment opportunity for Brazilian engineers outside the military. In 1874, as part of a series of educational reforms, the Escola Politécnica was established in Rio de Janeiro as a new engineering school fully separate from the military academy.

As historian Simon Schwartzman notes, "What gave meaning to the Escola Politécnica in Rio de Janeiro (as well as to the Escola de Minas and to some extent the Politécnica in São Paulo) was mostly their role in the creation of a new breed of elite intellectuals who could challenge the established wisdom of priests and lawyers in the name of modern science. The notion that society could be planned and ruled by engineers, which was well within the French tradition, would have a large impact in Brazil."[4] The creation of the Escola Politécnica was a triumph of the Instituto Politécnico Brasileiro, a group formed in 1862 to advocate a greater role for engineers in Brazilian society. Politécnicos, as they were called, viewed their profession as competing with lawyers to serve the state. For much of the nineteenth century, law school graduates (bacharéis) formed Brazil's intellectual elite and administrative class. The law schools of Recife and São Paulo (established in 1827) were the main institutions of higher education available to laymen, along with medical schools, and they were attended by sons of the landowning elite who had political ambitions. Politécnicos believed that engineers should replace the outmoded economic liberalism advocated by the bacharéis with a more protective national economic policy that would foster industrialization. As a means of promoting their profession's role in encouraging modernization, prominent engineers banded together in the Clube de Engenharia, founded in Rio in 1880; the club's statutes specified its link to "the development of industry in Brazil and the prosperity and co-hesion of two classes—engineers and industrialists."[5]

Politécnicos were among several groups of Brazilian professionals who embraced scientific theories of social progress during the First Republic. Many Brazilian doctors, lawyers, engineers, and military men adopted versions of Frenchman Auguste Comte's positivist philosophy and Englishman Herbert Spencer's evolutionary theory of societal change. Comte and Spencer each advocated increased national integration through improved transportation networks and industrialization, and these were key political goals of Brazil's growing urban elite in the decades surrounding the republic's formation. Both philosophies approached modernization as a process that should be guided by an educated middle class and an expanded government bureaucracy. Brazil's engineers and other urban professionals believed that a rational society required the leadership of scientifically trained men. They saw themselves—and not churchmen, landowners, or the untutored masses—as the appropriate agents of Brazil's transformation from a rural slave society to a modern democracy of free, increasingly industrial workers.

This posture was dramatically demonstrated during the 1889 coup that overthrew Emperor Dom Pedro II and launched the First Republic. Military officers, most of whom had trained as engineers, led the coup, and they hoped to bring an end to the dominance of rural landowners and their law-school-educated sons on Brazil's political life. Few officers came from the uppermost echelons of Brazilian society, and military promotions did not depend on the family-based patronage networks necessary for advancement in law and politics. Within a few years of the republic's formation, however, few military leaders held political office; some feared that official involvement in partisan negotiations would sully their profession's reputation for dispassionate, rational management. Military officers' reluctance to remain in political office allowed bacharéis, allied to the coffee barons and other landowners who still dominated national politics, to resume their prominence in public life. By the 1910s, many engineers and other middle-class professionals were disillusioned with the decadence and self-interest of the federalist republic's provincial oligarchies—especially the coffee growers and exporters in southern states.

In 1921, B. Piquet Carneiro, an engineer who had headed several IOCS departments, published a scathing critique of corruption in the republic's bureaucracy. Carneiro asserted that nepotism and mismanagement demoralized Brazil's public servants, including its engineers, and the Ministry of

Transportation and Public Works, responsible for the drought inspectorate, assigned its administrative posts based on political favoritism rather than competence. This, he posited, led to shoddy projects that reflected poorly on Brazil's engineering profession. The ministry accomplished little to further the national good, despite having a budget equal to one-third of the nation's revenue. Carneiro had long advocated building dam and irrigation works to aid drought victims in the sertão, but he felt that IOCS was not able to accomplish much because of administrative incompetence.[6] His disgust resonates with historians' assessment of the republic as a disappointment for many who had supported the 1889 coup. As Emilia Viotti da Costa writes, "The traditional rural oligarchy had been supplanted by a new one: the coffee planters of the [paulista] West and their allies, who, once in power, promoted only those institutional changes that were necessary to satisfy their own needs. Nov. 15 [1889] was thus a journée des dupes for all the other social groups who had hoped that the republic would represent a break with tradition."[7]

Engineers did not succeed in attaining an influential place within Brazil's federal bureaucracy until the 1930s administration of President Getúlio Vargas, and their subordinate political position during IOCS's early decades is one explanation for the agency's inability to significantly transform the sertão. Additionally, the personal experience and education of IOCS's managers did not provide the necessary cultural and intellectual perspective to assess the region's problems in social terms. Most engineers were conditioned by their exposure to Spencerian views of social evolution to think of society as a unified organism that could be guided along a well-defined path toward modernization. The drought agency's goal was overall regional improvement, and its Rio-based administrators did not consider that some social sectors might benefit from their efforts more than others—or at the expense of others. Such a limited view of regional development allowed upper-level IOCS staff to focus on measurable parameters, such as the volume of water in reservoirs or the number of hectares available for irrigated cultivation, rather than trying to assess the political factors underlying many sertanejos' misery during droughts. As in the Indonesian development projects analyzed by Tania Li, the agency's civil engineers successfully rendered drought a purely technical problem.[8] Their attention to technical issues like dam construction and increased water retention pleased IOCS's patrons in the national legislature, which was essential for the drought agency's survival given southern Brazilians' reluctance to devote federal funds to the sertão.

## Miguel Lisboa as IOCS's First Director

Miguel Arrojado Lisboa, a graduate of the Escola de Minas in Ouro Preto, Minas Gerais, directed IOCS from its founding in January 1909 until August 1912, and was reappointed in December 1920 by Epitácio Pessoa, remaining in the position until March 1927. Lisboa saw both climate and economic organization as important influences on sertanejo society; he believed that the uncertain climate affected sertanejos' character, making them resistant to planning ahead. He even credited climatic instability with encouraging the fanaticism often attributed to sertanejos, since people's imaginations could not be put to more productive uses while they waited to see what the seasons would bring.[9] But Lisboa also thought that the sertão's feudal land and labor organization adversely affected its economy, discouraging farmworkers from making improvements on property to which they had no secure title. He was optimistic about "islands of economic and social activity" where democratic ideals appeared to flourish. These were mainly in the mountainous areas of the sertão that had been settled since the 1870s by drought migrants seeking more fertile land. In such scattered pockets, like the Carirí Mountains of Ceará, settlers adopted cooperative irrigation arrangements to channel stream water to their fields.[10]

Lisboa, who was not a native Nordestino, made grandiose claims about the importance of IOCS's work for Brazil's national development. "The problem of drought is, in its highest expression, the very problem of our national integration," he wrote after concluding his first term as drought inspector. Geographic forces conspired to fragment Brazil into a mosaic of regions, each with its own distinctive identity. By helping to overcome impediments to economic progress in one large and important region, Lisboa hoped to help foster greater equality among all Brazilians. He described the sertanejo as "one of the most potent, latent forces in the country. But he is not sufficiently equipped to enjoy the improvements necessary for maintaining his existence and furthering progress on his land."[11] Educational programs were needed to help sertanejos make effective use of the projects launched by IOCS on their behalf, which included improvements to public health, irrigation, industry, cattle breeding, storage silos, wells, and reservoirs. Lisboa clearly saw the drought agency as having tremendous potential to alter living conditions in the sertão.

When IOCS was established, several theories were in vogue to explain what caused drought in the sertão. The scant climate data available for the northeast made it difficult to argue conclusively for one theory over another, though European meteorological studies were often cited to support conjectures about the sertão's weather patterns. The "rainmaker thesis" that had been popular since the mid-nineteenth century still had adherents who sketched ambitious plans to alter the climate by means of large reservoirs or massive reforestation. The most impressive of these was a proposal by Luiz Mariano de Barros Fournier to build a gigantic reservoir with a perimeter of 335 kilometers that would create a 400,000-hectare (1,544-square-mile) lake. Fournier expected the lake to dramatically increase condensation over a large area and make the normally dry winds blowing across the sertão wet. The reservoir would have been exceedingly expensive to build, and even advocates of substantial reservoir construction in the sertão regarded Fournier's plan with skepticism.[12]

More modest proposals for altering the sertão's climate focused on changing human behavior—in particular, ending deforestation. Since the early nineteenth century many observers had associated northeastern droughts with the profligate elimination of forest cover. A nuanced version of this theory, articulated by cleric and scholar Florentino Barbosa, argued that the northeast's interior plain was a fragile climate requiring careful adaptation by its inhabitants. Yet settlers since colonial times had not adapted to the limitations of their semiarid landscape, so human and natural forces operated in tandem to adversely affect the environment: deforestation increased wind erosion of the sertão's shallow soils, and abandonment of farms during droughts exposed the vulnerable land to intense sun and, later, heavy rains.[13]

During its initial years, IOCS gathered rainfall and river flow data for the sertão that spawned more detailed theories about what caused the drought cycle. Studies by Gilbert T. Walker, director of British India's meteorological service, emphasized the importance of the "southern oscillation" for northeastern Brazil. He demonstrated that Ceará's climate cycle could be correlated with those of other global regions affected by this mammoth atmospheric pressure pattern, making it somewhat possible to predict the onset of drought.[14] Sampaio Ferraz, director of Brazil's meteorological service, cited Walker frequently when discussing the influence of wind elevation and wind speeds on rainfall in the northeast. He argued vehemently for the importance of monitoring the sertão's climate via an expanded network of meteorological stations, which would improve the government's ability to predict droughts and respond before they became humanitarian crises.[15]

Lisboa combined the various climate theories available for the sertão into a multipronged strategy for reducing the adverse effects of drought. The primary influence on the sertão's rainfall pattern, he believed (following meteorologist Oswaldo Weber), was wind elevation: when ocean winds passed high over the drought zone, this inhibited condensation of atmospheric vapor, resulting in little precipitation. This dynamic was obviously not susceptible to human interference. But the construction of numerous reservoirs throughout the sertão, along with investment in reforestation to regulate runoff during rainy seasons, could reduce the impact of dry years. Lisboa advocated creating large reservoirs to allow for "full-scale irrigation and intensive cultivation" despite the sertão's unreliable rainfall.[16] He felt strongly that IOCS's priority should be to improve agriculture for local sustenance and export, since that was the drought zone's most important economic sector.

Lisboa also argued that the drought problem had to be considered from numerous angles—geographic, geologic, climatologic, botanic, engineering, hygienic, economic, and social—and that different solutions would be appropriate in different subregions. Small reservoirs served an important purpose in places where farmers planted in their basins once the water had evaporated—a practice known as *vazante* (ebb tide) agriculture. Wells were appropriate in areas like the state of Piauí, which had substantial subterranean water reserves overlain by permeable rock but lacked the broad plains necessary for large reservoirs. Ceará had many obvious locations for reservoirs of varying sizes: broad basins surrounded by irrigable plains situated in river valleys with suitable dam sites. Lisboa also recommended irrigating the banks of the São Francisco River—the northeast's most important waterway, and the sertão's only perennial river—with the aid of hydraulic pumps driven by water descending from high interior plains. As the economy of the river valley improved, the irrigated area could be expanded by adding more pumps. The drought inspector cautioned that transporting the São Francisco's water to other parts of the sertão, as had been proposed since the Great Drought, was unjust to those already living along the river's banks.[17] This would also be expensive and inefficient due to the distances that the water had to traverse, the high rate of evaporation in the sertão's hot and dry climate, and the need to pump water up significant elevations at some points.

As the first drought inspector, Lisboa confronted two competing approaches for sertão development. The option requiring the greatest disruption to existing land-use patterns would create irrigated small-farmer settlements on the banks of reservoirs. A more conservative approach involved storing water in reservoirs and wells for current landowners to sustain their production, particularly of cattle (which died in droves during droughts). This was often accomplished via "cooperative construction" on private or municipal lands, in which the cost of construction was shared by the drought inspectorate and landowners or local governments. The owner was theoretically obligated to provide water for workers and neighbors during drought years. Applications for construction on private land had to justify the reservoir's economic usefulness for ranching or agriculture. Most private reservoirs were justified as providing water for cattle and enabling farming in their evaporated basins during dry seasons. Few proposals for cooperative construction during IOCS's early years mentioned irrigated farming as a goal.[18]

Lisboa decided to begin with the more conservative strategy and move toward the more transformative one. He assumed that the extension of irrigation canals from agency reservoirs would evolve naturally despite initial opposition from rural dons. "What is most important is to satisfy the immediate aspirations of the Northeast; irrigation through canals will come as an inevitable consequence. In its own time, it will become an extreme political necessity," the drought inspector surmised.[19] Lisboa viewed irrigation as a long-term goal that would provide relief to vulnerable sertanejos during droughts, increase smallholding, generate revenue (from water use) to repay the cost of construction, and eventually contribute to the drought agency's coffers. What he did not consider was that IOCS's early work would increase the influence of rural coronéis who funneled federal drought aid through their patronage networks. IOCS could not ignore regional landowners' preferences, since the federal government had little muscle in the sertão without their blessing. Coronéis had no interest in seeing their workers settled on independent small farms, which would decrease their clients' dependence on patrons and thereby diminish the coronéis' political influence. Thus, building reservoirs on private land—despite being criticized as a federal subsidy for the wealthy—came to absorb the majority of drought-related aid.

Lisboa's enthusiasm for irrigation, and the opposition that he encountered from elites and the intended beneficiaries of such projects, is evident in documents concerning the construction of irrigation canals around Cedro Dam near Quixadá (190 kilometers south of Fortaleza). This was the first manmade reservoir to be constructed in the sertão; it was begun during the last decade of the empire in response to the horrors of the 1870s drought and completed in 1906. Irrigation canals, the construction of which was overseen by Carneiro, absorbed 30 percent of the project's budget. In 1910, Lisboa wrote to his superior, the minister of transportation and public works, that land surrounding the Cedro Reservoir should be expropriated and rented in smallholdings to needy sertanejos. Thus far landowners, who had not paid the government for the benefits provided on their property by the reservoir and canals, had strongly resisted such expropriation. Cattle herds belonging to these *proprietários* often damaged the government's irrigation works. Men like the mayor (*prefeito*) of Quixadá "speak falsely in the name of the poor population," Lisboa asserted, and did not in fact represent the poor; they were entirely motivated by self-interest.[20] He believed that a successful irrigated smallholder colony surrounding IOCS's first reservoir would demonstrate the potential economic and social utility for the sertão of the drought agency's works.

By 1913 Cedro supported over one thousand *vazante* cultivators within the reservoir basin as well as numerous fishermen. Lisboa and his subordinate engineers had also developed a detailed plan for administering Cedro's irrigation network to water at least 120 additional hectares (about three hundred acres) of land nearby. They proposed levying a tax on all owners of land served by the canals, since without this imposed cost many farmers would not bother to take advantage of the irrigation system's potential economic benefits (except during droughts). Judging by the series of exclamation points jotted in the margins of Lisboa's missive, it seems that his superiors thought this too extreme.[21]

As would be the case with many of the drought agency's irrigation works, property owners surrounding Cedro frustrated IOCS engineers due to their ignorance of and lack of interest in irrigated farming. They frequently allowed livestock to roam property (their own and others') traversed by irrigation canals, resulting in substantial damages that were expensive for IOCS to repair. In hopes of addressing this problem, drought agency personnel established the Escola Prática de Agricultura (Practical Agricultural School) in 1914, though its organizer noted grimly that he had already been trying to improve the administration of Cedro's irrigation system for several years, but to no avail.[22] In 1915 severe drought presented an opportunity to undertake canal repairs as a means of employing six thousand migrants who arrived at Cedro looking for work and food rations. During this crisis, irrigation water was provided at no cost to anyone willing to use it, to help *os indigentes* support themselves. Fish from the reservoir were also freely available. The number of *irrigantes* doubled, to 120 irrigating six hundred hectares of land (with holdings ranging in size from one-tenth of a hectare to forty hectares, averaging five hectares per irrigante).[23] Still, Cedro's managing engineer noted that his irrigantes required more oversight and training in this unfamiliar method of farming than he could provide. No one in the region had sufficient expertise to assist this effort (for example, by stipulating the volume of water necessary per hectare to grow different kinds of crops).

Two years later the number of irrigantes around Cedro had fallen to seventy, and property owners who benefited from irrigation canals were still not charged for use of the reservoir's water. The author of Cedro's 1917 annual report cautioned that desirable lots near the reservoir were not always reserved to aid the very poor, as they should be.[24] In 1919 another severe drought provided opportunity to repair blocked drains and damaged irrigation canals as a means of enrolling migrants who arrived daily in search of food aid.[25] The local mayor sent an urgent telegram to the drought inspector in Rio, pleading that Cedro's staff be authorized to provide more aid to the famished hundreds streaming to the reservoir site. They had been authorized to "recruit" seven hundred workers and had already enrolled twelve hundred, but children and the elderly perished daily from starvation in the encampments surrounding Cedro, and epidemics threatened to overwhelm this "martyred" population.[26] Forty irrigantes farming near the reservoir who were being charged a per-hectare rent by IOCS petitioned the head of its irrigation service to waive this fee given the aid that they themselves were providing to hundreds of families camped around their property and their own need to remain productive in the face of the drought.[27]

IOCS's experience at Cedro, its first attempt to move from reservoir construction to irrigation works and agricultural extension, indicates that engineers' positivist vision for what would benefit the sertão was contested by both wealthy landowners and marginal sertanejos. While landowners' resistance to the establishment of smallholder colonies is easy to comprehend given the links between land ownership, wealth and political power— and has frequently been cited as a reason for the drought agency's failure to adequately aid poor sertanejos during its first half century—archival records reveal that the sertanejo poor themselves were often not eager to adopt more intensive and regimented farming practices to improve their own long-term economic security. Drought agency engineers, trained in more industrial states like Bahia, Minas Gerais, and Rio de Janeiro, embraced a set of assumptions about the desirability of change guided by scientific and administrative elites that was culturally foreign to most sertanejos—both the elite and the very poor.

· · · · · ·

During Lisboa's first term, IOCS's budget increased by fifteen times (from 446:671$400 in 1909 to 6.686:227$100 in 1912). The agency used its funds to conduct surveys of northeastern botany, geology, climate, rainfall, and river flow in order to establish a solid basis for dam construction. This was a response to the frequent accusation that earlier dams, particularly Cedro, had not repaid the investment made in them due to meager understanding of local hydrology. Many of the surveys were conducted by foreigners: Philip von Luetzelburg was an Austrian botanist who spent most of his career studying Brazilian flora;[28] Orville Derby was an American geologist who led the Serviço Geológico e Mineralógico do Brasil (Brazilian Geological and Mineralogical Service) in assessing the northeast's subterranean waterways and the São Francisco River system to prepare for well and reservoir construction; and a team of geologists from Stanford University—Roderic Crandall, Horatio L. Small, Ralph H. Sopper and Geraldo A. Waring—published a number of reports for IOCS between 1910 and 1923 on water sources in different subregions of the sertão. Crandall also worked under Derby to produce a detailed map of Ceará for the drought inspectorate showing town locations, rough elevation, waterways, telegraph lines, and existing and planned roads and railroads.[29]

IOCS also moved rapidly into reservoir construction. By September 1911 the agency had begun constructing three large reservoirs, five medium reservoirs (between fifteen hundred and eight thousand acre-feet capacity each), and three small reservoirs and had drilled sixteen wells—with numerous other reservoirs and wells planned.[30] The sertão's reservoirs generally had lower dams than reservoirs of similar capacity elsewhere in the world due to the shallow basins that regional geography offered; this made them cheaper to build (many were constructed from packed earth rather than masonry), but they were subject to a high rate of evaporation in the sertão's intense heat.

In 1912, following four months of work under Lisboa's direction, Waring published a review of IOCS's efforts and assessed the agency's achievements very favorably, comparing it to the U.S. Reclamation Service: "Since the organization of the Inspectorate two years ago the region has been mapped, the principal river basins have been examined and the feasible reservoir sites found, and general plans for the development of the region have been outlined. Besides these direct works, the Inspectorate has made studies of the agricultural capabilities of the various portions of the region, has established rain-gauging stations, and has initiated a systematic measurement of the discharge of the rivers."[31] Waring summarized IOCS's development strategy as follows, reflecting Lisboa's confidence that the agency would soon begin building smallholder irrigated settlements and spur significant social reorganization: "It is believed that the people will migrate to the irrigated areas during times of drought, and will remain there. Thus the present scattered population will become collected into agricultural communities [through the purchase of land with water rights by individuals], and a general advance will be made in the condition of the people and in the commercial development of the region. Provision will also be made for the [planned] colonization of some of the irrigated areas, in order that a more rapid development of the region may be aided by this means."[32] Waring concurred with Lisboa's belief that immigration to the sertão would increase once irrigation works were constructed. Both men hoped that an influx of industrious, experienced, and—significantly—white farmers from Europe and elsewhere in Brazil would provide native sertanejos with role models for productive farming. Waring perceived sertanejos, "of negro descent mixed with the native Indian and with Portuguese" to be somewhat lazy and ignorant, but he thought that they were held back as much by their environment as by any racial or cultural handicap since, "when the natural conditions permit, they improve their chances for betterment."[33]

Beginning in 1915, the engineer who would oversee IOCS's operations in Ceará as chief of the inspectorate's "first district" agreed that the technological improvements provided by the agency would have significant social impact. Thomaz Pompeu de Souza Brasil Sobrinho had, like Lisboa, trained at the Escola de Minas in Ouro Preto, and he was a member of the powerful Accioly family whose wealthy members held important political offices in Ceará. Brasil Sobrinho had suggested in 1912 that the material gains accruing to sertanejos as a result of reservoir and canal construction would create an atmosphere in which destabilizing political influences could not flourish: "In agricultural regions in which artificial irrigation assures the complete success of many branches of industry, one does not encounter strikes; there are no socialists or, even less, anarchists; the agriculturists do not even concern themselves with politics. It is sufficient that an administration provide them regularly with the most important element for their production—water for the cultivation of their fields."[34] A prominent *cearense* landowner himself, and an advocate of technological improvements to farming and ranching, Brasil Sobrinho portrayed IOCS's infrastructural investments as a means of ensuring social peace by enriching all residents of the drought zone through improved technology.

Brasil Sobrinho cautioned Lisboa that the federal government's efforts to combat drought had historically been short-sighted. Poor planning under the imperial government in the late nineteenth century contributed to a false but widespread national belief in "the uselessness of public action in relation to the drought problem."[35] He thus proposed that IOCS create a long-term, region-wide reservoir plan to foster irrigated cultivation and provide an economic anchor for the sertão. He was concerned that extractive industries on the rise in the region (such as the export of carnauba palm wax) would wreak havoc with its economy, as had just happened in the Amazon when its supply of high-quality rubber diminished.[36] Brasil Sobrinho commended Lisboa's comprehensive river basin analysis, which he hoped would put irrigation plans on a firmer footing than they had ever been in the past.[37] He believed that rents from farmers with reliable harvests and the increase in land values resulting from irrigation would repay the government's investment in these efforts.

Lisboa launched his administration of Brazil's federal drought agency with a strategy for addressing the climatic crisis that appealed to the sertão's ranchers and export farmers—namely, reservoir construction. He assumed that irrigated farming by smallholders would follow once reservoirs were built, diminishing the inequities that left many poor sertanejos, without food or cash reserves, vulnerable to droughts. In preparation for reservoir construction, Lisboa commissioned scientists to collect data on the sertão's physical environment. But IOCS never undertook a parallel analysis of the region's social organization. Lisboa and his engineering colleagues were confident that their improvements to hydrologic infrastructure would engender social progress without requiring deliberate meddling in political affairs. This technocratic confidence proved to be misplaced. IOCS became part of Nordestino elites' bureaucratic apparatus for controlling resources in the sertão, and instead of eroding landowners' power, the agency provided yet another mechanism through which rural patrons could dole out government largesse and intensify clients' dependence on them.

## Justifying Federal Investment in the Sertão during the First Republic

Federal funding for IOCS was a subject of intense discussion among members of the national legislature. Brazil's 1889 constitution gave its central government limited control over the states, and state autonomy was a cherished principle of the republic's organization. Politicians resistant to aiding the sertão argued that northeastern state governments should first attack the drought problem rigorously themselves. Supporters of federally funded drought works countered that the oligarchs who controlled the northeastern states were not interested in regional economic growth if it might reduce their influence, so the federal government needed to assume expanded responsibility on behalf of sertanejo citizens. More moderate voices noted that northeastern states did not have sufficient resources to counter the drought problem alone, and improving the region's infrastructure would benefit Brazil as a whole through expanded markets and tax revenue.

Advocates of federal drought aid worked to convince representatives in Brazil's national legislature that drought works were an appropriate and sound investment. The Municipal Council of Quixeramobim, Ceará, argued in a 1911 memo to the federal Ministry of Transportation and Public Works that reservoirs were a form of defense against an enemy, equivalent to the military defenses against foreign invaders that the federal government already funded. The differences between drought and war were that droughts caused greater devastation to a broader spectrum of the population, including women and children, and demoralized an entire region.[38] Other northeast boosters pointed to the mineral wealth of the sertão and its agricultural potential, which could not be fully realized without adequate water. Several supporters of federal investment in drought works claimed that outmigration of young men during drought years had hindered the northeast's development; one benefit of alleviating the periodic crisis would be retaining that productive population in the sertão. A parallel line of reasoning, following the decline in European immigration to Brazil during World War I, was that reducing the migration of Nordestinos to the south would help to justify subsidizing new European immigration to that more dynamic region.[39]

Other proponents of federal drought aid sought to combat the common national perception that sertanejos were lazy and unlikely to avail themselves of assistance. They portrayed sertanejos' contributions to the national economy as remarkable given the severe obstacles to economic productivity confronting them, including drought, disease, and poor transportation networks. Engineer Raymundo Pereira da Silva argued in a 1907 speech to Rio de Janeiro's prestigious Clube de Engenharia that sertanejos had been commendably resourceful considering the poor infrastructure available to them. "What might one expect of [the sertanejo] if such obstacles were removed or reduced?" he asked.[40] Since all northeastern cities were forced to cope with a periodic onslaught of desperate migrants from the interior, the entire region's development had been crippled by drought. Pereira da Silva contended that providing technological improvements analogous to those available in Brazil's wealthier regions would unleash the natural industriousness of sertanejos and reveal the latent fertility of their native land.

Many questions about increased federal authority over the sertão were answered by reference to irrigation programs in foreign countries. Such comparisons emphasized that federal investment in drought works did not violate the principle of state autonomy and that such expenditures served the national interest. The head of the Clube de Engenharia cited the United States as a nation deeply committed to states' rights that had nevertheless seen the benefits to be gained from placing irrigation and river navigation under federal authority: "In the great North American republic, where the autonomy of states and municipalities is so respected, a national agricultural irrigation service has been implemented, as has been done in India, Egypt and Europe. The American people place the national interest above all else and do not hesitate to confer such responsibility on their federal government. The country's prosperity depends in large part on the fulfillment of this federal obligation. Not only the irrigation service falls under the national government's jurisdiction, but interior navigation as well, which affects the nation just as much."[41] Supporters of federally funded sertão development noted the remarkable transformation of the American West made possible by the nationalization of interior waters as public property under federal jurisdiction. They recommended that the northeast's nonperennial rivers, which were the primary sites for reservoir construction, be nationalized in order to eliminate the legal complexities of administering dam projects.

Engineers who promoted northeastern drought works calculated that providing irrigation to sertanejo farmers would yield a net economic gain despite the substantial initial cost. They alleged that a federal irrigation program in Mendoza, Argentina, had stimulated the creation of new industries that drew on the products and revenue from increased agricultural production. This was a useful example, since that neighboring Latin American country was also a federation of states. But the three semiarid regions most commonly used to illustrate how irrigation could benefit the sertão economically and socially were Egypt's Nile River valley, India under British rule, and the American West. All such references were intended to emphasize that Brazil would be participating in a noble and progressive tradition of rational water management if it undertook extensive irrigation works in its drought zone.

The Nile was on the minds of many engineers working in the sertão at the turn of the twentieth century. British engineer J. J. Revy claimed in 1878 that his proposed reservoir at Lavras, Ceará, would create a valley as pro-ductive as the Nile had become under British administration.[42] Engineer Joanny Bouchardet observed in the 1910s that the English reaped rents of seventy-five million pounds—more than twice Brazil's federal budget—annually from the Nile's irrigated banks,[43] and he believed that the first dam along the Nile had altered rainfall patterns, since shortly after its con-struction the region enjoyed significant rain for the first time in years. Civil engineer J. A. de Castro Barbosa praised the efforts of Britain's Lord Cromer and Lord Dufferin to reorganize agriculture along the Nile and rescue in-habitants from the threat of droughts and floods, improving public health in the process: "Today along the river margins, alongside Pharaonic monu-ments, [the English] have erected modern constructions around which the people devote themselves to labor with the same ardor as their ancestors, but now peacefully, because they will no longer be cheated of their efforts by the invasion of floods or the occurrence of droughts."[44] Barbosa felt that the São Francisco, the northeast's most important perennial river, could eas-ily match the Nile and other great rivers around the world for navigational and agricultural importance.

Brasil Sobrinho compared the northeast's need for irrigated agriculture with India's, since large numbers of people died in both places as a result of drought and famine. He commended British administrators in India for managing simultaneously to aid the native population and generate signifi-cant income from agriculture, which he claimed approached, after just a decade, 90 percent of the amount that the colonial government had spent on irrigation works. Brasil Sobrinho quoted Britain's Lord Salisbury and Lord Cromer on the preeminent importance of irrigation to India's economic development.[45]

The U.S. government's willingness to invest vast sums irrigating its sparsely inhabited, semiarid West received frequent mention in early liter-ature promoting northeastern drought works. The U.S. Reclamation Service was viewed as having undertaken an even more Herculean task than IOCS attempted, since some of the areas that the United States tried to irrigate were truly desert (whereas the sertão enjoyed adequate rainfall much of the time). The Reclamation Service was reported to have transformed several western states into fertile, populous, and increasingly wealthy areas in a very brief period of time. Brasil Sobrinho reported to Lisboa that the Rec-lamation Service spent US$43 million in its first five years of operation, a "fabulous sum" that had nevertheless been wisely apportioned and resulted in twelve million people occupying fourteen million hectares of irrigated land around large reservoirs and canal systems.[46] He emphasized that the United States had undertaken such intensive development solely out of economic interest, without the added impetus of a starving population that should spur Brazil to more vigorous response.

As IOCS's first director, Lisboa was cautious about the applicability of foreign examples to the northeast. He thought that India was more analogous to the sertão than the United States was, because India's river margins had been heavily settled before canal and dam construction began. Parts of the sertão that received sufficient rain in most years—notably, in the states of Ceará and Paraíba—were relatively densely settled. These areas were subject to particularly acute crises during droughts, when large numbers of people were forced to migrate.[47] The sertão's existing population presented a particular challenge for Brazil's drought agency, since these inhabitants would require substantial education via agricultural extension programs in order to succeed as irrigation farmers. In Lisboa's understanding, the U.S. Reclamation Service had not confronted a significant native population whose way of life would be changed by irrigation works.[48] Lisboa clearly believed that development planning for the drought zone required consideration of social factors as well as environmental ones, though he did not emphasize the differential effects of drought on social classes within the sertão.

Aarão Reis followed Lisboa as drought inspector from 1913 to 1915 and, in response to the devastating 1915 drought, oversaw a reorganized commission, the Obras Novas Contra as Secas (New Works to Combat Drought) through 1919. An abolitionist and admirer of the United States, at least as Alexis de Tocqueville described it, Reis had graduated from Rio de Janeiro's Escola Politécnica in the 1870s and subsequently taught social sciences there. He was a prominent member of Rio's Clube de Engenharia and held numerous government positions prior to his work for IOCS, in railway expansion (where he collaborated with doctors Carlos Chagas, Oswaldo Cruz, and Belisário Penna on public hygiene efforts), electric and telegraph networks, and constructing the city of Belo Horizonte, Minas Gerais. Reis's training and early professional experiences made him a classic positivist: he saw well-administered public institutions and science as the primary engines of social progress.[49] Wary of dramatic social rupture, he believed that scientifically trained men should guide social evolution methodically, helping the state to provide education, medical care, and other services that would accelerate national economic development. In relation to the drought problem, Reis favored practical, technological solutions like expanding transportation networks to improve market access for sertanejo farmers, and he was dismissive of Lisboa's geologic and botanic studies as too re-moved from the urgent task of reducing the economic impact of drought. Reis's criticism of prior IOCS work and confidence that he could direct the inspectorate from its Rio de Janeiro headquarters led to tensions between his young staff, hired from Rio's Escola Politécnica, and experienced Nor-destino personnel like Brasil Sobrinho.[50]

· · · · · ·

Throughout the 1920s, advocates of northeastern drought works continued to compare IOCS's efforts to the irrigation and development of semiarid re-gions abroad. J. A. Fonseca Rodrigues, a paulista engineer, understood In-dia to have profited handsomely from investment in irrigation by 1919; Ceará, in contrast, had not. Most of Ceará's first-generation reservoirs were not large or deep enough to capture sufficient water. Cedro Dam, begun dur-ing the late empire and completed in 1906 after many delays, did have significant capacity, but the surrounding hydrology was poorly understood at the time the site was chosen, so its reservoir rarely filled higher than fif-teen meters out of a potential twenty-four. The suffering of sertanejos dur-ing droughts was exacerbated by Ceará's poor interior transportation network, Fonseca Rodrigues argued, whereas India benefited from the ex-tensive rail services constructed by its British colonizers.[51] Northeastern Brazil was in greater need of improved hydrologic and transportation in-frastructure than India, he concluded. It had yet to receive a fraction of the investment that the British had expended on a mere colony.

Fonseca Rodrigues was also impressed by the conservation efforts of the U.S. Bureau of Forestry and similar reforestation programs in England, Ger-many, and India since the turn of the century. In his view, the main point of increased tree cover in the sertão would be to alter rainwater flow and better secure farmable soil. Brazilian landowners were already awarded monetary compensation by the federal government for planting particular tree species, but the establishment of national forests would be a more ap-propriate way to ensure public gain from valuable trees.[52] The Indian and U.S. governments earned revenue from their forest reserves through the sale of timber and other products, and Brazil's federal government could follow the U.S. example of selling improved public lands to farmers and foresters, with revenue returning to a reclamation fund to make regional development self-financing. Fonseca Rodrigues believed that having alternate industries to ranching made good economic sense in the sertão given the substantial challenges that ranchers faced to sustain livestock through droughts. Transforming ranchers into irrigation farmers would require significant training and possibly the example of new immigrants to demonstrate unfamiliar farming techniques, so he advised exploring a variety of new industries.

As was true during IOCS's first decade, the activities of the U.S. Bureau of Reclamation continued to receive close attention in the 1920s from supporters of northeastern drought works. Ildefonso Albano, a former mayor of Fortaleza (and future governor) who represented Ceará in the national legislature, used Reclamation Service data to illustrate how profitable the rent or sale of irrigated land could be. He noted that the U.S. government had invested $116 million over fifteen years to make a sterile desert productive—though without the added pressure of cyclic tragedy that the sertão bore.[53] Brasil Sobrinho thought Nordestinos should feel ashamed for not implementing simple irrigation technologies that "semibarbaric" peoples had used for centuries, given the tremendous human cost of drought in their region.[54] He upheld white settlement in the American West as a prime example of the social and economic gains to be made from irrigation.

Another view of the sertão's potential for social transformation through hydrologic infrastructure came from Dwight P. Robinson, president of the American engineering firm contracted by the Brazilian government in 1921 to construct the Orós Dam. (Due to technical difficulties resulting in part from insufficient initial surveys of the chosen site, the dam was not completed until 1960). This was a massive undertaking in Ceará's Jaguaribe River valley; the resulting reservoir was to hold over three billion cubic meters of water and have an irrigation capacity of several hundred thousand hectares. As a result of the project's size and predicted cost, it received significant criticism. Robinson wrote to Lisboa in August 1923, responding to a critique of the Orós venture published by one "Dr. Moraes e Barros" in the *Jornal do Brasil*.[55] Robinson's analysis—self-interested, since his own firm's contract was at stake—directly confronted the racial biases underlying many Brazilians' skepticism of the drought agency's mission to transform the sertão through improved infrastructure. Moraes e Barros deemed sertanejos to be "undersized, malformed, and unintelligent; lacking initiative and energy," but Robinson's own staff, "highly experienced with many classes of labor," had received quite a different impression during their two years working in Ceará and Paraíba. The American engineer described sertanejo men as disciplinable workers with " bulldog persistence" that served them well through years of drought and pestilence. They were, he allowed, intellectually children, but this was the result of circumstance: due to their constant preoccupation with obtaining sufficient food, they had never attained much formal education. Moraes e Barros recommended "bringing European colonists to Ceará" for the region to prosper; but in Robinson's opinion the Europeans, inexperienced in that environment, would fare "worse than the natives." Many of Robinson's engineers had joined his company after years with the U.S. Bureau of Reclamation. That agency's investment in irrigated agriculture, he asserted (backing his argument with copious economic data), had born substantial fruit within the first fifteen years, even without a starving population to spur government action.[56]

· · · · · ·

Advocates of the drought agency's technocratic mission during Brazil's First Republic were somewhat misled by their optimistic assessment of similar undertakings in Egypt, India, and the United States. Many recent histories take a more critical view of irrigation's achievements in those places. Timothy Mitchell links Egyptian dams and irrigation canals to mosquito-borne epidemics, dependence on foreign-manufactured fertilizers, and malnutrition when fertilizer could not be obtained.[57] He and other historians note the difficulty of trying to predict how much land could be irrigated by a particular river system and the occasionally dire consequences of miscalculation. Historians of India observe that many irrigated settlements did not achieve the "yeoman farmer" ideal embraced by British colonial authorities, since members of dominant castes who were granted land often hired poorer farmers to work for them.[58] Elizabeth Whitcombe has emphasized the deleterious health and environmental effects of British India's irrigation works.[59] The proliferation of irrigation canals and railroad embankments increased the prevalence of stagnant water pools in areas where malaria's mosquito vector already bred profusely.[60] Drainage was impeded during heavy monsoon rains, causing salts to accumulate near the soil surface and crystallize during dry seasons.

Environmental historians Donald Worster and Marc Reisner have pointed out the undemocratic nature of irrigated agriculture in the western United States, since federal reclamation efforts often subsidized cultivation by wealthy farmers. Donald Pisani views the Reclamation Service as an ideological failure that aided ranchers more than family farmers. According to Pisani's analysis, powerful state interests exercised their influence in the U.S. Senate to direct the agency's resources to projects that served their own ends. Land speculation and the difficulty many settlers faced making a living within reclamation projects led to the concentration of substantial land holdings in a few hands. Other historians have observed that insufficient climatic and hydrologic data for subregions of the American West caused many irrigation efforts to fail. Long-term ecological damage, including that resulting from soil salinization, was not foreseen or adequately managed.[61]

Such recent histories of irrigation in Egypt, India, and the United States indicate the misperceptions of many proponents of drought aid in northeastern Brazil who saw those regions as models of rational technological development. But there is a more fundamental problem with the insistence by IOCS's staff that Brazil's sertão would follow the historical trajectories of distant semiarid lands. The agency's managing engineers and their political allies relied on analogies to foreign development projects without carefully examining the particular context and challenges of the sertão—especially its political institutions and social dynamics. Most of the bureaucrats who promoted IOCS's drought works clung to the belief that transferring technological solutions from places like the American West would be sufficient to re-create the Nordeste in the image of those societies. This conviction reflected engineers' positivist assumption that society was moldable through rational, scientific means. IOCS's directors were confident that the prosperity they saw (somewhat unrealistically) accruing to North America's Western settlers in particular was fundamentally the result of expanded technological infrastructure and not of the broader political and legal culture in which those systems functioned. Civil engineers' faith in narrowly technological solutions to the drought problem aligned with their desire to be seen as essential agents of transformation for a region that many viewed as chronically backward.

### Navigating a Middle Politics on the Front Lines of Drought Aid

In contrast to many of their superiors, drought agency *técnicos* who managed reservoir projects in the sertão often developed an understanding of the drought problem that focused on factors beyond climate. Confronting desperate poverty and devastating epidemics, many site managers came to view drought crises as resulting from power imbalances and inequality as much as from any natural factors. Telegrams sent by engineers and agronomists working in the sertão indicate the predicament of technical staff who, faced with the extreme misery of their compatriots, began to comprehend the social complexity of the sertão's drought problem. These scientists worked at an uncomfortable intersection where technology's tantalizing promise of a "magic bullet" to end poverty without political conflict collided with the stark likelihood that only direct confrontation with regional elites could reduce the vulnerability of the sertanejo poor. As middle-class agents of social change in a region characterized by profound inequality, IOCS's project managers often advocated on behalf of their desperate workforce—and in opposition to their own superiors and local elites—while simultaneously battling the perceived ignorance and recalcitrance of the people whom they strove to help. Civil engineers' positivist vision for what would benefit the sertão was contested from both above and below, and this led them to carve out a political middle ground (what Michael Ervin in the context of revolutionary Mexico terms a "middle politics") that acknowledged some grievances voiced by marginal sertanejos while ultimately promoting the technocratic agenda for regional development most desired by landowning elites.[62]

Consider the Forquilha Reservoir in Ceará. It was first proposed by businessmen in the Sobral area during the 1915 drought as a means of sustaining hundreds of starving sertanejos while bringing federal funds and new infrastructure to their region. These *comerciantes* were inspired to send telegrams to President Brás, they claimed, in response to the agony that they witnessed among thousands of starving retirantes who flocked to their city in search of food. Private charity had been exhausted, and they were now relying on national patriotism and humanitarian sentiment. Fielding this request, the drought agency's director responded that preliminary studies of the proposed reservoir had not been undertaken, but a twenty-five-kilometer road linking Sobral to Meruca could commence as a means of employing many migrants and removing this burden (and presumably the fear of anarchy) from Sobral's citizens.[63] Within a few years, IOCS had also made progress on the proposed dam site. Staff reached "friendly" (that is, uncontested) indemnification agreements with several homeowners in Campo Novo, a small town located within the planned reservoir basin. In order to determine property values, engineers evaluated house size, style and construction material (e.g., tile or thatch roof), the length of fencing, and existing plantings. One man who owned a large house, several smaller homes, nine thatched huts, and crop storage facilities received 25 percent of the agency's indemnification funds—an indication of his substantial wealth (and influence) relative to neighboring residents.[64] Other homeowners began to protest their indemnification offers after the local bishop complained that the town church had been unjustly indemnified at half its market value. "It would be an act of justice to indemnify [Campo Novo residents'] property for its real worth," site manager Abelardo dos Santos (a civil engineer who had graduated from Bahia's Escola Politécnica in 1911) appealed to his superiors—particularly since in most cases the small houses and land were all that these families owned.[65] Yet indemnification funds arrived from Rio de Janeiro only early in 1928, when property owners around the reservoir sent urgent telegrams to the drought inspector explaining that the reservoir was filling but they still had no resources with which to relocate their homes, families, and livestock.[66]

Construction of the Forquilha Dam under the guidance of Santos sustained many laborers through the 1919 drought. A letter from inspector Aarão Reis to the minister of transportation and public works indicates the ethical machinations engaged in by drought agency managers when fulfilling their professional obligations. Within a few lines, Reis describes the suffering of drought migrants around Forquilha as "distressing . . . due to the drought scourge" and as "an opportunity to take advantage of the large number of retirantes already in the construction area."[67] The drought inspectorate provided humanitarian aid to people whom it pitied, yet it exploited this desperation to serve its own need for cheap labor. Funds and supplies for construction sites were chronically late. Santos's telegrams to his superiors in Rio de Janeiro repeatedly request more buckets, shovels, pickaxes, and wheelbarrows—and always increasing "credit" to enroll hundreds more men and boys in the federal project (men were paid 1$400–1$800 daily, depending on the task assigned to them, and boys were paid $500–1$200).[68]

Santos's regular communications from the sertão to Rio de Janeiro provide detailed accounts of the suffering of retirantes who arrived at the Forquilha site, and on several occasions he negotiated with his superiors on their behalf. In his July 1919 report to the drought inspector, Santos tallied 1,246 workers under his management who were providing succor to 8,722 people around the dam site (an average of seven family members per worker, typical of other agency projects). Hundreds more arrived in search of work each day.[69] Since cash was scarce, as always, Santos arranged to pay his workers in "points" (credits) that could be accepted by any of several private food suppliers. He invented this system to avoid the company store dynamic that often arose when construction workers were obliged to purchase from a single fornecedor. Even so, he told the drought inspector, prices of necessities like beans and *rapadura* (blocks of raw sugar) continued to rise as stores throughout the region diminished, and the agency's established daily wages were insufficient to sustain large families. In his subsequent monthly report, Santos requested permission to give workers earning the lowest diárias (less than 2$000) additional half points on Sunday (their one day off), because otherwise they were too weak from undernourishment to work productively on Monday.[70] A flu epidemic was raging through the reservoir's encampment, and Santos's September report notes with little fanfare that 10 percent of the migrant population surrounding the Forquilha site (140 people) had died of the illness that month—the disease's impact having been compounded by malnutrition; he had contracted a doctor and was providing medicines to the sick. He concluded grimly that the combination of drought and illness had severely reduced his workers' productivity so that their actual construction accomplishments that month were modest.[71]

Under these desperate circumstances, Santos briefly broke with IOCS practice and provided aid to the families of men who had died of the flu, even when there was no one to replace that deceased worker (and thereby earn the household's food rations). Upon hearing this, inspector Aarão Reis sent a telegram to Santos asking him to cease this unorthodox practice.[72] Reis gave Santos special permission to pay enlisted workmen who were seriously ill in order to aid their recovery (since without the daily wage they had no means of obtaining food), but if their flu proved fatal, the men's bereaved families must be instructed to seek help elsewhere unless another family member (such as a teenage son) could be enrolled to replace the original worker.[73] As Santos well knew, there was nowhere else those destitute families could turn.

By January 1920, with more than thirteen hundred manual laborers still enrolled at the dam site (in addition to thirty more specialized personnel, such as office administrators and electricians), Santos's workers included about two hundred women. This was unusual, and seems likely to have been a compromise reached in response to his request to continue sustaining families of male workers who had died of the flu. Following his superiors' orders, Santos limited female workers' maximum daily wage to half that of the lowest diária allocated to grown men, or 1$200;[74] this provision was to account for their presumed lesser strength and reduced productivity. Given that Santos already deemed the male workers' minimum diária to be below subsistence level for a family, one can assume that the circumstances of families with only a woman to enroll were extremely precarious.[75] Forquilha's midlevel administrative staff, such as office clerks, earned eight times the average diária of female workers.

Santos's communications with the drought inspector by letter and telegram were always brief and empirical, with little narrative embellishment. Yet one senses his growing fatigue as the months of 1920 rolled by with little reprieve. He requested permission to open additional construction sites for which preliminary studies had already been conducted so that he could enlist as many of the drought refugees streaming daily into Forquilha as possible. By May the number of enlisted workers at three proximate sites for which he was responsible mounted to over 2,500, providing food aid for more than fourteen thousand retirantes. In order to justify enrolling more starving families, Santos and his colleagues often deliberately lowered the already modest technological level of their construction processes—for instance, having the workers haul dirt in buckets rather than wheelbarrows. If the Rio de Janeiro office insisted that every family receiving food rations must include one drought works employee, then the engineers facing legions of starving families were determined to find ways of employing as many "heads of household" as they could. Even with the first return of rain, sertanejo farmers weakened by months of drought could not return home immediately, as there was nothing there to nourish them until the next harvest. Some site managers who found ways to enlist desperate migrants until sufficient rains returned may have been motivated by fear of mob violence as much as by sympathy—a few telegrams to the Rio headquarters mention this serious concern.[76]

In a close analysis of the Tucunduba Dam site records, historian Aline Silva Lima portrays Santos as a highly dedicated, but—in the view of his superiors—somewhat naive young man who envisioned IOCS transforming sertanejo culture in numerous respects. Santos saw the sertão's notorious nepotism as an obstacle to regional modernization, and he vigorously opposed clientelist networks.[77] This provoked the wrath of fornecedores at Tucunduba who had, prior to Santos's arrival, also served as labor recruiters, securing jobs for their loyal clients. Santos defended his lowliest workers against exploitation by fornecedores, publishing tables of acceptable prices for various goods based on surveys of prices charged in nearby market towns, away from the desperate refugee encampments. Because of these efforts to eliminate the patronage and profit opportunities rampant at many dam sites, Santos and his staff were occasionally threatened with violence and his equipment was subject to vandalism.

IOCS's construction works clearly altered local social and economic landscapes. They presented a new form of federal patronage and a chance to break with existing clientelist networks, offering a range of jobs with salaries allocated based on skill levels, including machine mechanic, office administrator (available to literate sertanejos), and various trade occupations like carpenter and mason. Linda Lewin attributes an increase in banditry by the late 1920s to sertanejos' unwillingness to return to "subjection" by landlords after they had become accustomed to working on federal dam- and road-building projects.[78] The construction sites included public health stations and (if more than fifty workers were employed) schools. Santos established a school for the practical instruction of reservoir laborers' families that he tried to require their children to attend.[79] He felt that he was engaged in moral reformation of the sertão, targeting both the corruption of local elites (such as exploitative fornecedores) and the recalcitrance of agricultural workers who did not strive to improve their families' and communities' economic circumstances through more rationalized labor regimes. Santos strove to import a version of factory management to the sertão, which he understood as essential to the transformation that sertanejos must undergo as part of their region's economic and social modernization; he aimed to inculcate in his workers a commitment to efficient production and a daily rhythm dictated by the clock.[80] Like other positivist drought agency *técnicos*, Santos wished to imbue his workers with a more progressive mind-set, one less resigned to letting nature's caprice determine their fate.

## Debating Drought Works during Epitácio Pessoa's Presidency

Advocates of northeastern drought aid found a champion in Brazil's first Nordestino president, elected in 1919 as a "compromise" candidate acceptable to power brokers in the most influential southern states. Epitácio Pessoa, scion of a prominent Paraíban family, governed Brazil from June 1919 to November 1922. He had attended law school in Recife from 1882 to 1886 and was elected representative to the national legislature for Paraíba in the early 1890s at age twenty-six. He served in several positions during the early years of the republic, including minister of justice and the interior from 1898 to 1901, *procurador* (attorney general) from 1902 to 1905, and minister of the Supremo Tribunal Federal do Brasil (equivalent to the supreme court) from 1902 to 1912. In 1916 he was elected as a senator from Paraíba, and he was reelected to the senate following his term as president. Throughout his career in national politics, Pessoa remained closely tied to Paraíban interests.[81]

President Pessoa was adamant that northeastern drought works had both an economic and a humanitarian justification. He frequently emphasized that the drought crisis should be understood as a national rather than regional problem and that his attention to it reflected his Brazilian patriotism, not his "northern soul."[82] Pessoa questioned the motives of those who objected to sending federal funds to his native region, asking, "Why is it only when one speaks of the Northeast that certain patriots' itch for economy becomes aroused?"[83] His administration spent more than double on Brazil's three most populous southern states what it spent on the eight-state Nordeste region, yet it was criticized primarily for its expenditures on northeastern projects. Following his presidency, Pessoa defended his investment in drought aid as "the honest payment of a debt of honor from the nation, which could not continue to be indifferent to the periodic sacrifice of so many lives and the criminal abandonment of so many resources."[84] He believed that irrigation would cost Brazil less than what the U.S. Reclamation Service spent per hectare and promised to generate comparable revenues; cotton yields might easily be twice as high per hectare as in Egypt or the United States.[85] When improved ranching and increased electrical generation were added to the rise in agricultural output and land values that Pessoa anticipated would result from irrigation, the economic justification for investing in the sertão seemed clear.

In 1919 Pessoa renamed IOCS the Inspetoria Federal de Obras Contra as Secas (IFOCS; Federal Inspectorate for Works to Combat Droughts) and dramatically increased its funding (which had fallen during World War I, when a sharp decline in coffee exports reduced federal revenue). His first budget allocated to IFOCS almost five times the amount that Lisboa had been granted as drought inspector in 1912. In 1921–22 Pessoa more than doubled this and apportioned a similar amount for railroads and ports in the northeast. Altogether during his presidency, Pessoa directed 15 percent of federal revenue to drought works. In the same period the legislature permanently committed 2 percent of national revenue to a fund for drought aid, to be matched by 2–5 percent of revenues from each state in the drought zone (payable once in donated land, thereafter in cash). The federal government would have administrative and use rights over drought works until the cost of constructing them had been repaid, via sale or rent of irrigated plots and an irrigation and drought works conservation tax.

Pessoa viewed his investment in drought works as a continuation of IOCS's initial research and construction projects, and at the end of 1920 he reappointed Lisboa to direct the agency. Lisboa embarked on an ambitious program of reservoir construction, picking up where he had left off at the end of his first term as drought inspector. He hired three foreign companies to oversee dam projects: British firms Norton Griffiths & Co. and C. W. Walker & Co., and U.S. firm Dwight P. Robinson & Co, whose senior administrators had directed major projects for the U.S. Reclamation Service. In defense of this selection Pessoa insisted that the use of foreign technicians reflected Brazilian engineers' lack of experience with large-scale reservoir building but not any lack of skill; indeed, national firms had been invited to bid on IFOCS contracts but had declined after understanding their scope.[86] In addition to undertaking new reservoir projects, Lisboa asked scientists to investigate improvements to the northeast's cattle industry.[87]

The substantial increase in IFOCS's funding under Pessoa raised the stakes in national debates about the drought agency's agenda and accomplishments. Critics cited inadequate achievements during IOCS's first decade. Even many northeast boosters were frustrated by the agency's paltry accomplishments. Cearense representative Ildefonso Albano chastised IOCS for having been so ineffective during its first five years that only the grisly 1915 drought justified its continued existence to a skeptical public. In Albano's provocative metaphor, the northeast suffered from a chronic malady and had been granted the temporary relief of a *curandeiro* (folk healer), not an effective cure by a scientific doctor.[88] IFOCS administrator Brasil Sobrinho also rebuked the agency for its modest progress, pointing out that political bickering and "the strange prestige accorded to certain ignorant and authoritarian functionaries" routinely undermined the good intentions of many government bureaucrats in the republic, including his colleagues. As a result, "nothing, or almost nothing, was accomplished of real and practical utility" during the drought agency's first decade.[89]

Widespread acknowledgment of the drought agency's inadequate performance became ammunition for opponents of federal investment in the sertão during the 1920s. In response, Albano accused his colleagues in the national legislature of remaining willfully oblivious to his native region's recurrent crisis. He described the drought scourge to them vividly, calling it "the gravest and most urgent socioeconomic problem in Brazil"—though one lacking attention because it did not impinge upon the welfare of the country's most powerful people.[90] Sertanejos were not begging for charity, he insisted; they wanted to be free of the curse that had prevented them from improving their lot over generations. Brasil Sobrinho was also adamant that northeastern states required well-managed federal assistance to help them cope with droughts. Waiting for those states to finance development efforts on their own would create a vicious cycle of inaction, since only irrigation could bring Ceará and its neighbors sufficient wealth to do that. Brasil Sobrinho justified aiding the sertão as a sound national investment based on the northeast's existing contributions to the federal treasury, which he claimed were higher per capita than the national average despite crippling droughts.

As one of a growing number of influential cearenses committed to industrialization and economic diversification, Brasil Sobrinho strove to promote Ceará's overall economic growth, blithely confident that this would aid the very poor as well as the more fortunate. He supported awarding grants for reservoir construction on private lands, despite the fact that this directed the benefits of IFOCS's work primarily to the stratum of cearense society least desperate for aid. This stance was certainly pragmatic. Northeastern states were limited in what they could—or were willing to—fund outside their capitals (which themselves were inadequately supplied with basic public services and infrastructure). Partnerships between state government and private interests, or federal subsidies for privately funded construction, were a way to accomplish what the states could not manage on their own and to demonstrate the potential gains to be made from further public investment.

Even so, Brasil Sobrinho recommended that IFOCS pursue various ways of reducing the impact of drought beyond the reservoir construction that had occupied much of the agency's first decade. Large reservoirs should be linked to irrigation canals as the surest means to provide an economic boost to the sertão and alleviate hunger during droughts. He also advised the agency to undertake secondary projects such as improved cattle breeding, adoption of crops better adapted to the climate, and increased mechanization of agricultural production. Brasil Sobrinho viewed reforestation (especially on sandy soils where reservoirs could not be constructed) as another worthwhile project that would allow economic diversification through timbering and fruit cultivation. He hoped that wise investment in Ceará's irrigable land would spawn a population increase, further boosting the state's productivity. Brasil Sobrinho understood that many Nordestino politicians pressured IFOCS to direct federal patronage to their clients, which resulted in widely scattered projects. Yet the agency would do better to focus on densely populated areas where drought's human costs were highest, encouraging cultivation of cotton and of secondary crops like sugarcane and tobacco in those places.[91]

Brasil Sobrinho accused IFOCS's staff of being largely inexperienced in agronomic matters—which, he believed, were as important to the success of sertão development as engineering expertise. "From this result embarrassing disasters," he lamented; "inappropriate works, poorly studied, poorly planned, poorly constructed, and poorly chosen."[92] In general, he thought IFOCS should leave the administration of drought works to people familiar with the sertão's environmental conditions. He warned that most sertanejos did not understand the dramatic measures necessary to change the circumstances handed to them by nature, due principally to "the religious education that we have which tends toward coddling in the [human] spirit a backward fatalism."[93] Well-educated *técnicos* knowledgeable about the sertão needed to make wise decisions about how to deploy available resources to counteract the drought.

In response to the varied criticisms leveled against IFOCS early in his presidency, Pessoa established a special commission to study the drought problem. It was headed by two members of the national legislature, Ildefonso Albano of Ceará and Cincinato Braga of São Paulo. In the commission's report Albano emphasized the need to ensure that irrigation benefit potential drought migrants more than large landholders. He proposed that dam construction be preceded by land expropriation so that existing landowners would not be unduly enriched by federal investment. As a concession, they could retain 10 percent of their property (minimally, twenty hectares) that would benefit from the new reservoir. Albano also recommended that control of drought works devolve to local irrigators' associations rather than to state governments once the federal government recouped its expenses through water usage fees and land taxes. These measures aimed to democratize control over water.[94]

Paulista representative Braga proposed a very different development strategy for the northeast, focusing on the region's transportation infrastructure rather than its water resources. In his view, improved railroad lines were a more defensible federal investment than reservoirs, since they served a larger number of Brazilians: roads and rail lines were useful in all seasons, while reservoirs were critical primarily during droughts. Braga described roads and rails as "works whose practical, political, humanitarian and economic results can be predicted with much more certainty and whose success is much more assured than that of any other potential solution."[95] The total cost of improved transportation for the Nordeste (assuming 1,500 kilometers of roads and 3,200 kilometers of rails) should be less than two-thirds of the estimates for Pessoa's reservoir plan (or about 200,000 contos), he contended.

Braga enumerated various problems with the reservoir strategy: evaporation and soil salinity were known challenges in the sertão's equatorial climate; dam breaks had occurred even in countries with a more highly trained engineering corps (he referenced the 1889 flood in Johnstown, Pennsylvania); federal expropriation of land for reservoirs was not clearly constitutional; and the rise in land values predicted to follow irrigation might simply cause inflation rather than yielding real increases in wealth. Braga questioned the justice of displacing hundreds of households in order to create reservoirs that would primarily benefit a small number of farmers and ranchers. He also felt that cultivating cotton, one of the primary commercial crops intended to be grown along reservoir margins, was risky. Domestic consumption of the fiber was modest, and there was no reason to believe Brazilian cotton would do well in foreign markets.

Braga argued that the federal government needed to focus its economic development efforts on national industrialization, technical education, and manufacturing for export—a set of priorities that, unsurprisingly, particularly suited São Paulo's interests. His analysis of Brazil's economy and existing deficit indicated that public coffers were in no condition to finance uncertain, expensive projects.[96] (As reward for his criticism of inflation and deficit spending, Braga was named first director of the Bank of Brazil in 1923 by Pessoa's successor, President Artur Bernardes.) Braga asserted that Brazil's engineers deliberately underestimated the cost of public works in order to get their projects approved; in the sertão, cost estimates for transporting materials to construction sites, land expropriation, and interest on borrowed capital were routinely left out of proposed budgets.

Braga encountered opposition from several quarters. Brasil Sobrinho argued that rail lines had thus far provided more benefit to the northeast than planned irrigation only because the hydrology and soil conditions around early reservoir sites had been poorly analyzed prior to construction, and many proved unable to support irrigation on a large scale. Penna expressed concern about the adverse effect of rail lines on local ecology and public health; he cautioned that railway construction required deforestation, and embankments would block water channels, contributing to the formation of stagnant pools where disease-carrying mosquitos could breed. Inland rail lines encouraged settlers to abandon their farms for uncultivated territory farther west, where pioneers would be exposed to new diseases (and often remained only a short while, migrating again when epidemics struck). For railroads to anchor productive regional development, Penna argued, they would have to be accompanied by sanitation projects around new settlements. Yet such efforts could more profitably be undertaken to stabilize rural populations where they already lived, eliminating the immediate need to run expensive rail lines farther into the interior.[97] Penna noted that the financial woes Braga cited had come about despite substantial national investment in railroads, so those investments did not necessarily generate positive economic outcomes for the nation.

In response to the above observations, Braga advised that if IFOCS did continue to focus on reservoirs, for greatest efficiency the agency should concentrate these in a few areas to which most inhabitants of the sertão could easily migrate. This recommendation, similar to Brasil Sobrinho's a few years earlier, met with concerted opposition. Naturally, many regional politicians preferred to see new infrastructure distributed throughout the sertão, to ensure that their constituents would benefit directly. But proponents of constructing reservoirs across the interior also recognized sertanejos' profound unwillingness to relocate, except when faced with acute crop shortages. This sentiment was noted by a second commission sent by Pessoa in 1922 to assess IFOCS's progress. That commission's report emphasized sertanejos' persistent desire to return home even after they had experienced easier living conditions in the mountains and on the coasts as a result of drought migration. Regional planners who attempted to move sertanejo farmers to new irrigated colonies around reservoirs often found themselves at odds with the farmers they intended to help.

Based on the variety of recommendations for northeastern development that he received, Pessoa adopted a multipronged strategy comprising reservoir construction, reforestation (which, he believed, would produce climate change), and increased road and rail networks. He anticipated that irrigation networks would eventually extend from the large reservoirs, but few canals were constructed during his presidency. Drawing on Braga's proposal, Pessoa agreed that improved transportation was essential to IFOCS's regional development plan. Better roads would help deliver construction materials and relief provisions when needed and would bring settlers to new farming areas. "It is an absolute fact that the backwardness of our interior is due exclusively to the difficulty of transportation [through the sertão]," Pessoa asserted hyperbolically in one interview.[98]

Pessoa saw irrigated agriculture as among the most significant goals of his investments in the northeast. Yet the majority of funds spent on the sertão during his administration were for reservoirs, wells, and improved transportation infrastructure. Cincinato Braga warned Pessoa that his increased expenditures on drought works would result in a discontinuation of IFOCS's projects by the next administration. Indeed, Pessoa's successor, Artur Bernardes, eliminated the "permanent" federal drought works fund in 1923 and halted construction in the sertão by 1925, leaving IFOCS with a skeletal budget. The agency was allotted 3.826:749$300, compared to the 145.947:350$000 that it had enjoyed three years earlier—a decrease of over 95 percent. Bernardes's priority was to balance Brazil's budget after Pessoa's inflationary spending. Nonetheless, he invested substantially in Southern Brazil during his tenure. A native of Minas Gerais, Bernardes exhibited little concern for the problems of the Nordeste or the plight of impoverished sertanejos, and there was no severe drought during his presidency to spark renewed national interest. Although Miguel Lisboa remained drought inspector until 1927, he could accomplish little after Pessoa's term ended.

Reviewing Brazil's twentieth-century regional development efforts, economist Albert Hirschman surmised that Pessoa's bold attempt to "irrigate the desert" in a span of several years was a gambit intended to counteract the likely reduction in funds by his successor. Pessoa hoped to generate sufficient momentum to ensure that even an opposing administration would be compelled to complete his major drought works, and he chastised legislators who recommended more gradual development in the sertão: "Always the idea of postponement, in a country where discontinuity in administrative measures is a characteristic feature of every government!"[99] As it turned out, this pattern applied to his own cherished projects.

### Pessoa's Frustrated Legacy in the Drought Zone

In 1922, toward the end of his presidency, Pessoa dispatched a second commission to report on IFOCS's progress, hoping to rebut journalists' accusations that the agency was wasteful. He nominated three southerners "of integrity and above suspicion" to conduct the survey and convince "the sane part of the country" that drought works were "a more than justifiable sacrifice imposed on the treasury."[100] These dignitaries included General Cândido Rondon, who had directed the extension of telegraph lines from Rio de Janeiro westward into the Amazon basin; I. Simões Lopes, Pessoa's minister of agriculture; and Paulo de Moraes Barros, a prominent paulista politician. Rondon and Lopes had trained as engineers and participated in establishing the republic in 1889; Barros was a physician and businessman from an influential family.[101]

The three men published their findings in the *Diário Oficial* (the organ of the national legislature) and the *Revista Brasileira de Engenharia* "in order to make the nation's great endeavors [in regard to the drought] publicly known, and to justify the [financial] burden they entail through clear explanation of their ends and of the advantages and benefits they will bring us."[102] The report generally praised the northeast works: "All the necessary materials for the ports, reservoirs, railroads, and roads appear to be in perfect order and well looked after. All the studies of hydrology and hydrography and of irrigation, as well as the furnishing of materials from Brazil and the transportation of imported supplies, are taken charge of and perfectly executed by

IFOCS."[103] The commissioners described the agency's professional staff as "well-prepared, proficient, diligent, and disciplined." They were concerned that IFOCS's work had frequently been misjudged out of ignorance and had been unjustly condemned in the national press. Rondon, Lopes, and Barros deemed the government's central goal for drought works to be exactly what they would recommend—namely, securing sertanejo farmers in their home region. They characterized the sertão's population as "[g]enuinely national—an amalgamation of cross-breeding among whites, blacks, and aborigines in every shade of intermixture [sub-mestiçagem], without a collective predominance of any particular type—possessing latent energetic qualities and notable resilience, despite the degenerative factors assailing them."[104] Yet despite their relatively high opinion of sertanejos' productive potential, they advocated establishing colonies of immigrant farmers from southern Europe to serve as an example for native farmers and combat the "indifference and depressed spirit" that a lifetime of hardship had bred.

The commissioners emphasized the importance of stabilizing food cultivation by constructing small and medium reservoirs (without irrigation networks) throughout the drought zone. This seemed already to have been accomplished in the three states that they visited: Ceará, Paraíba, and Rio Grande do Norte. They recommended a secondary focus on large dams for economic development to make the sertão's major rivers perennial and provide irrigation for 160,000 hectares (roughly 620 square miles) along riverbanks. That area, representing about 1 percent of the drought zone's 156,000-square-kilometer expanse, was all that the inspectors judged to be irrigable without considerable expenditure (to pump water long distances or to great heights). Nonetheless, they believed that the resulting increase in cotton production would be sufficiently profitable to justify building the necessary irrigation networks. They did, however, criticize IFOCS's decision to construct wide roads suitable for automobiles when oxcart trails could often serve local needs adequately at one-twentieth the cost. Pessoa replied that reliable, paved roads were essential to transport construction and relief supplies rapidly to the interior.

The commission's tally of works under way or completed at the time of their survey included:

556 kilometers of new roads (estradas de rodagem) in use
1,887 kilometers of roads partly usable and 144 kilometers under
    construction
1,193 kilometers of cart trails (caminhos carroçáveis) in use

786 kilometers of cart trails partly usable

292 kilometers of new railway in use and 445 kilometers more under
    way

229 reservoirs constructed, reconstructed, or under construction,
    holding 339,777,983 cubic meters of water

thirty-seven reservoirs planned, five in the planning stage, and
    fourteen more projected

139 tubular wells drilled

various port works, a telephone network, and a geographic service

Pessoa noted that this list represented only a portion of IFOCS's achievements through 1922. His commissioners had not visited all of the states in
which the agency was active, nor had they accounted for every type of project undertaken. The total cost for all of the work completed by IFOCS
through 1922 was 304,040 contos, Pessoa asserted—less than the half million or more reported by his critics. Of this, 187,770 contos were for importation of materials sufficient to complete all of the projects that were
launched during Pessoa's administration, and 33,527 contos addressed regional needs beyond the drought per se.[105]

By the time the commissioners' report was published, Pessoa's presidency
had ended. During Bernardes's term in office, Pessoa debated his northeastern legacy publicly with the three men he had asked to survey his administration's accomplishments in the sertão. Their primary point of disagreement
concerned how many acres would be irrigable if IFOCS's planned reservoirs
were completed. Rondon, Lopes, and Barros felt that the hydrologic and topo-
graphic data available for the sertão were insufficient to precisely determine
irrigable areas. Many reservoirs had been built in places where the surrounding land was not farmable; in other cases reservoirs held too little water to
irrigate all of the cultivable land around them. The commissioners also
thought that Pessoa's estimates for water capture did not account for the high
level of evaporation in the sertão or for the absorption of runoff by the soil,
both of which reduced the volume of rainfall that could be retained. Additionally, they questioned Pessoa's estimates for how many hectares of water
were required to irrigate specific crops.[106] In disputing Pessoa's defense of his
legacy in the sertão, Rondon, Lopes, and Barros cautioned that some of the
former president's predictions were based on studies conducted in foreign
semiarid regions and that local observations were essential to sound planning. They implied that Pessoa's projections about the likely economic value
of IFOCS's projects were often exaggerated.[107]

Pessoa responded that in the years prior to his presidency, Brazilians had clamored for concrete action in the sertão, not just hydrologic studies. Accumulating further data about the region was therefore not politically feasible. Even so, he asserted, existing surveys (for which the most reliable data was available at the inspectorate's Rio de Janeiro headquarters, not at the offices in the northeast where commissioners had obtained their information) were more than sufficient to direct IFOCS's work—especially when used in combination with information from foreign irrigation projects. In fact, Pessoa argued, scientific data available for the northeast were more extensive and reliable than data for many foreign regions that had already launched successful irrigation programs.[108] Data for the volume of water needed to irrigate a variety of crops had been reliably arrived at in foreign studies under conditions identical to those pertaining in the northeast, he claimed. These figures were all lower than those used by the commissioners, and in some places the commission's estimates of irrigable acreage had already proven to be too conservative. As an example, Pessoa cited the Piranhas River system in Paraíba, where the area irrigated by the end of 1923 was already 50 percent larger than what the commissioners had thought possible.[109]

Much of the debate between Pessoa and his three-member commission focused on the Orós Dam in Ceará. Orós was IFOCS's most ambitious undertaking up to that point, though it would not actually be completed until 1960. It was planned to have a capacity of 3.5-billion cubic meters, which would have made it the largest reservoir in the world in the 1920s. Rondon, Lopes, and Barros thought that Pessoa's estimate of 122,500 irrigable hectares around Orós was double the likely area. They believed that reliable irrigation estimates had to allow for the possibility of a three-year drought. But the former president felt that irrigation estimates should be arrived at for good years, to indicate the potential increase in annual productivity that reservoirs offered. Disagreement over how to predict a reservoir's usefulness for irrigation was not resolved in the prolonged debate between Pessoa and his critics. In 1927, Brasil Sobrinho added his weight to the Orós Dam controversy, noting that the amount of water needed to irrigate each hectare could probably be reduced by half, but only if farmers were provided with substantial education in intensive cultivation methods. In this way, he believed, 122,500 hectares around Orós could perhaps be irrigated, as Pessoa claimed.[110]

Pessoa defended himself in numerous newspaper articles and in his 1925 memoir, *Pela Verdade* (For the truth).[111] He aimed to counter accusations that funds sent to the northeast on his watch had been misspent. Rio de Janeiro's business-oriented *Jornal do Comércio* supported Pessoa's claims and clarified that most of the difficult preliminary work of dam building, such as transporting equipment and excavating foundations, had been accomplished before President Bernardes cut funding to IFOCS. The sudden paralysis of IFOCS by Pessoa's successor caused many nearly completed dams and roads to deteriorate in subsequent years.

· · · · · ·

The three members of Pessoa's drought commission were not the only people to take issue with the former president's defense of his administration's accomplishments in the drought zone. Ursulino Dantas Veloso, an agronomist who had worked for IFOCS in Ceará from 1920 to 1924, wrote decades later that the inspectorate's reservoir plans in that period had been too technologically ambitious. Dantas Veloso felt that Lisboa should have devoted more resources to drilling wells, providing cisterns to households, building small and medium earthen dams, and assisting with soil conservation along reservoir banks. The result of Lisboa's premature launch of more complex and expensive projects was that his works lacked sufficient analysis and political support, and most reservoirs were never accompanied by the irrigation canals necessary to make them useful for farming. Dantas Veloso feared that large reservoirs encouraged sertanejos to settle in a few areas, whereas the sertão's geography and fragile ecology made it more sensible to stabilize the population where it had originally scattered, throughout the drought zone.[112] His criticisms raise important questions about the professional and political priorities of IFOCS's managing engineers that led them to focus on large dams and reservoirs rather than more modest hydrologic or farming projects.

Adopting a different line of criticism, engineer E. Souza Brandão published numerous newspaper articles during Pessoa's presidency promoting education and public health as essential elements of sertão development overlooked by IFOCS. Souza Brandão felt that none of the reservoirs or road networks proposed by his colleagues in the drought agency could achieve their broadest goals for social improvement without better popular education and general sanitation. Aware of the low level of formal education throughout Brazil, and particularly in the northeast, Souza Brandão encouraged a secular, practical approach to schooling that would include nutrition, child care, and agricultural production as standard subjects. He argued that the physical health of sertanejos required measures to reduce hookworm, syphilis, malaria, tuberculosis, and various animal-borne diseases—all of which drained workers' productivity.[113]

At the end of the 1920s the man who followed Lisboa as drought inspector, José Palhano de Jesus, summarized his recommendations for future drought works in a memo to his successor. Palhano de Jesus explained that his agency's work had stagnated due to confusion about land rights and ownership of reservoirs, along with a lack of funds for irrigation systems. Only Quixadá Dam in Ceará was ready to support irrigated cultivation. He recommended that all but the largest reservoirs be temporarily rented to municipalities or farmers' cooperatives to raise funds for irrigation, with use rights returning to the federal government for emergency purposes during droughts. He also emphasized that once irrigation works were completed, planned colonization along reservoir banks, accompanied by agricultural extension programs, was essential to reap their full benefit.[114] Palhano de Jesus's concerns indicate what little progress IFOCS had made in promoting irrigated smallholding during its first twenty years, despite the conviction of Lisboa, Pessoa, and others that such projects were essential for averting future drought crises.

Pessoa believed that irrigated farming could markedly improve the sertão's economy and reduce sertanejos' vulnerability to droughts. Yet in defending the drought agency's accomplishments during his administration, Pessoa concentrated on narrow technical disputes, such as how many hectares of a particular crop could be irrigated by a given volume of water. In the twilight of Brazil's First Republic, most discussions of IFOCS's work continued to focus on questions of hydrologic infrastructure: how many dams and canals should be constructed, where they should be located, and what their potential to increase agricultural output would be. The need for improved agricultural instruction and general education for small farmers, as well as public health services, were mentioned occasionally. More controversial issues like land expropriation to increase smallholding, the manipulation of the sertanejo electorate by coronéis, and the concentration of political and economic power within a narrow social stratum were not addressed by the politicians and administrators who had the greatest influence over the drought agency. Partly as a result of this limited, technocratic focus, Pessoa's commitment of substantial funds for sertão development did little to aid the majority of sertanejos when the next major droughts arrived in the 1930s.

## Degrees of Vulnerability and the Technocratic Lens

To encourage investment in their states during the federalist First Republic, northeastern elites argued that droughts impeded national progress and increased regional disparities. Their pleas for aid ignored the social stratification within the sertão that contributed to the suffering caused by droughts. Instead, propagandists for federal drought relief portrayed the sertão's population as uniformly in need of assistance. As is the case in many such "natural disasters," sertanejos were not in fact equally affected by the climatic scourge;[115] some were much more vulnerable than others when droughts struck, and some were partly responsible for the precarious circumstances of their compatriots. Tenant farmers' lack of secure work contracts, personal capital, and crop surpluses made them particularly vulnerable to climate fluctuations and resulting harvest failure.

Epitácio Pessoa's presidency presented an opportunity to direct increased federal aid to the sertão, since he was genuinely concerned about his native region. But the president from Paraíba did not pursue policies that would have jeopardized landowners' dominion over the sertão's human and natural resources, such as establishing a mechanism for land expropriation around reservoirs. This reluctance to oppose the interests of his own political network and social class limited Pessoa's ability to increase ordinary sertanejos' security in the face of unpredictable droughts. Although he viewed irrigation networks as central to economic and social progress, few were constructed during his administration. The reservoirs that IFOCS did undertake suffered from the subsequent administration's neglect.

The civil engineers who oversaw IFOCS during Brazil's First Republic were caught between the self-interest of Nordestino elites and the self-righteous skepticism of southern politicians who claimed that the sertão was too poorly governed for federal aid to be used wisely there. In order to maintain northeastern elites' support for the drought agency within the national legislature, IFOCS's managing engineers aligned their priorities with those of regional power brokers. These engineers firmly believed that providing the sertão with improved roads, reservoirs, and (eventually) irrigation networks would lay the groundwork for both economic expansion and social evolution. Yet the technological systems that they constructed often reinforced elite control over critical resources in the sertão.

IFOCS's staff mistakenly viewed drought as the sertão's defining feature and compared the sertão's development to that of other semiarid regions in the world. Yet in a profound sense, drought was not the sertão's central problem. As Belisário Penna and Arthur Neiva surmised during their health survey of the northeastern hinterland in 1912, widespread poverty and inequality were the core causes of economic stagnation in the region. Penna repeatedly wrote that in order to help the sertão, Brazil's federal government needed to focus not only on disease but also on the political factors contributing to sertanejos' malnourishment, which made their bodies easy prey for opportunistic microbes. Similarly, to reduce the drought calamity for the sertão's landless and smallholding poor, the federal inspectorate needed to focus not only on the semiarid climate but also on the region's landholding and labor organization. This alternate perspective began to be adopted by agency technocrats who managed dam sites, as this chapter has demonstrated. In the 1930s, agronomists newly hired by the drought agency began to articulate a similar "middle politics" in relation to regional development, negotiating between their own technocratic vision, sertanejo farmers' resistance to intensive cultivation, and elite opposition to social change.[116] This will be the focus of chapters 4 and 5.

It was of course professionally expedient for IFOCS's engineers to identify roads and reservoirs as the central means of mitigating the sertão's recurrent calamity. If problems of social organization were at the heart of drought crises, then engineers would have had a less significant role to play in resolving them. By rendering drought a technical problem of insufficient infrastructure, as they were well trained to do, civil engineers assured their profession's centrality in developing solutions to it.[117] Yet as long as IFOCS declined to address the sertão's glaring social inequities, it had little chance of curing the region's afflictions.

**Patronizing the Northeast**

IFOCS under Vargas in the 1930s

· · · · · · · · · · · · · · · · · · · · · · · · · · · · · · · · · · · · · · · · · · ·

In the final years of Brazil's First Republic, a widely acclaimed novel presented a very different picture of sertanejos than that offered a quarter century earlier by Euclides da Cunha. *A Bagaceira* (1928) was written by José Américo Almeida, a thirty-year-old native of Paraíba state. Almeida subsequently headed the federal Ministério de Viação e Obras Públicas (Ministry of Transportation and Public Works)—which had jurisdiction over the federal drought agency—twice under President Getúlio Vargas. *A Bagaceira* was a watershed in Brazilian literary history because of its evocative use of regional vocabulary and customs and its focus on harsh circumstances not usually depicted in Brazilian literature. It depicts *retirantes* (drought migrants) who enter a sugar plantation in search of food and work during the 1898 drought. The title roughly translates as "trash," but more accurately as "bagasse," which refers to the pile of used sugarcane stalks discarded near sugar mills, and figuratively to the people (such as menial laborers) who were discarded in such places.

The images of retirantes passing by the sugar mill are probably based on Almeida's own childhood memories or other eyewitness descriptions of drought survivors:

> Emaciated ghosts, their shaky, unsteady steps seemed like a dance as they dragged themselves along in the manner of one who is carrying his legs instead of being carried by them. They walked slowly, looking back behind them as if anxious to return. There was no hurry to arrive, for none knew where he was going. Expelled from their paradise by swords of flame, tormented by furies, they wandered aimlessly on, fleeing the sun, their guide in this enforced nomadism. . . .
>
> They were more dead than alive. Life, urgent life, showed only in the eyes, whose pupils reflected the all-consuming sun. . . . Their knees were bent . . . ground down by fatigue.[1]

In other passages, Almeida vividly details the devastation wrought by the drought on parched crops, cattle, landscapes, and people.

Almeida's protagonists are racially and culturally distinct from da Cunha's primitive *mestiços*. He describes them as light-skinned former landowners, of European descent and admirable breeding. Almeida contrasts these sertanejos, men of honor and women of robust health, with the lazy and lascivious *brejeiros* (a pejorative term for people of African descent from Paraíba's *agreste* region—located between the semiarid sertão and the coast) who resent their presence. The retirantes' determination and self-reliance in comparison to the unmotivated plantation workers with whom the drought forces them to associate is couched in overtly racial terms.[2] Such sertanejo prejudice against the descendants of slaves who work the sugar mills is a common theme in the folk tales and popular poetry of northeastern Brazil. The view of blacks as inferior is evident in stories that describe the mixed-race marriages of retirante daughters as an affront to their parents, for whom sugarcane workers represented servility.[3]

In the end, the months spent at the plantation ruin the retirantes. A young girl becomes involved in an illicit affair with the mill owner, and her father is imprisoned for trying to avenge his daughter's virtue. Years later, the mill owner's educated son defends the girl's aging father as a victim of societal irresponsibility:

> Who is the more guilty—the prisoner who killed one man, or society, which has, through criminal neglect, allowed thousands to die?
> Before he was accused the prisoner was himself the victim of society. The droughts come at regular intervals. Everyone was able to predict when the catastrophes would occur. But the authorities did nothing to prevent them. They never tried to overcome the vicissitudes of nature which at one moment gives generously, the next takes all. Even today the uncertainty of fruitful labor in the [sertão] is still waiting for some rational measures to improve conditions and give security to the sertanejo.[4]

This is the lesson of Almeida's book, that sertanejos merit compassion and assistance, not ridicule and neglect. The drought migrant's lack of options leads to his own moral and material downfall and the degeneration of northeastern society overall.

As head of the ministry responsible for the Inspetaria Federal de Obras Contra as Secas (IFOCS; Federal Inspectorate for Works to Combat Droughts) during the early 1930s, Almeida tried to redirect the agency's priorities to measures that he believed would stave off the excruciating suffering he had witnessed as a child. During a severe drought in 1931–32, Almeida and drought inspector Luiz Vieira were widely praised for their dedication to aiding victims at refugee camps and worksites. But President Vargas's interest in the sertão's woes was largely motivated by the political capital such attention could bring him, and it diminished (relative to urban industrialization and other priorities) once the crisis abated. As a vocal defender of the northeast's interests within national politics, Almeida became disillusioned with Vargas's fickleness toward his native region and resigned from the ministry in 1934. Civil engineers remained in charge of drought works, despite Almeida's addition of an agricultural service to IFOCS in 1932. Due to the continued dominance of an engineering mind-set that favored reservoir construction as the solution to recurrent drought, Almeida's promotion of irrigated smallholder cultivation as a development priority had little impact during the 1930s.

## Vargas's Paternalism and Tempered Reform

During the 1920s, oligarchic power was in decline throughout Brazil as the country's economic and social organization increased in complexity. Opposition to the influential coffee-exporters in southern states grew among members of the country's middle and upper classes who felt shut out of negotiations between power brokers in Minas Gerais and São Paulo. A group of young military lieutenants (*tenentes*) demanded reforms on behalf of Brazil's growing middle class, including economic modernization, honest governance, and national unity under a strong central government. The global economic crash of 1929 and resulting crisis in the coffee market further reduced the power of Brazil's rural oligarchs. In 1930 several allied factions overthrew the federal government and put Vargas in the presidential palace. A member of the ranching elite from the state of Rio Grande do Sul, Vargas was a populist leader with greatest support and influence in the south. From the start of his administration (which lasted until 1945 and then resumed with his reelection in 1951), Vargas pursued national industrialization and administrative centralization.

When Vargas first came to power it was important for his modernizing agenda that Nordestinos be regarded as important potential contributors to the national economy. Vargas had displaced an elected president from São Paulo, and his administration was strongly opposed by *paulista* elites; the state rebelled against his government in 1932. The new president thus

Workers at the São Gonçalo agricultural post, Paraíba, illustrating sertanejo racial types with predominantly European, African, and indigenous characteristics (according to agronomist José Guimarães Duque). Source: *Boletim da Inspetoria Federal de Obras Contra as Secas* 11, no. 2 (1939): n.p.

needed to downplay the importance of São Paulo to national development. As part of this strategy, Vargas emphasized the role that the Nordeste must play in accelerating Brazil's productivity. To this end, Vargas's "culture managers" sought to dispel earlier depictions of the region and its inhabitants as backward and racially degenerate.[5] His Departamento da Imprensa e Propaganda (Department of Press and Propaganda) upheld the *caboclo*—a person of mixed Indian, white, and (less emphatically) African descent, commonly from the interior northeast—as the guardian of traditional Brazilian values. This nostalgic image was in pointed contrast to the German and Italian immigrants who had been settling in São Paulo and the southeastern states since the mid-nineteenth century. Vargas's propagandists tried to renovate the national image of Nordestinos by describing them as "whitenable," in the *branqueamento* tradition embraced by elites since the 1880s.

As Jerry Dávila explains, whiteness was understood in the 1930s as a social category implying health, virtue, modernity, and an escape from (black) degeneracy.[6] Stanley Blake finds that the Vargas government portrayed sertanejos in particular as "non-white, non-black, non-rebellious, capable and educable"—and thus fit subjects for development projects that aimed to incorporate them into the national economy.[7] Engineer Thomaz Pompeu de Souza Brasil Sobrinho, from Ceará's division of the federal drought agency, published two articles on the *homem do nordeste* (northeastern man) in 1934. These downplayed the importance of African elements in the population and emphasized sertanejos' native (Indian) heritage as ennobling—the root of their tenacity and thirst for autonomy.[8] In this portrayal, sertanejos were important potential contributors to Vargas's modernizing agenda, requiring only guidance from Brazil's more educated citizens. IFOCS offered one means to accelerate their "whitening" through technical education.

Vargas's several administrations were rife with contradiction. He came to power under the mantle of a revolution committed to reforming the self-interested policies of the republic's oligarchs, yet to preserve his own authority he found it necessary to placate elites in the regions that had been dominant during the preceding decades. Vargas's corporatist approach to national governance offered moderate reforms, guided and overseen by state bureaucracy, as a concession to urban workers. More extreme transformative agendas, such as those allied to communist ideology, were co-opted or suppressed. Vargas sought gradual social progress through conciliatory strategies that addressed the most acute concerns of Brazil's urban workers without unduly threatening elite interests. Historians have criticized Vargas's efforts to aid Brazil's poor for their paternalism—particularly his promotion of state mediation between workers and industrialists in place of independent labor unions.[9] Scholars also note that Vargas's policies intended to benefit the poor were effective only in major cities. Nevertheless, the president's populist style garnered the support of workers and the poor throughout Brazil. Vargas oversaw significant federal expansion, which helped him to secure middle-class support as well.

The dramatic political shifts that shaped Vargas's several administrations indicate the substantial challenge of promoting economic and social modernization without succumbing to political fragmentation. Following a revolt by *paulistas* in 1932, Vargas allowed the republic's old guard greater power. In the mid-1930s, a political party viewed by many as communist gained strength with a platform of achieving the unrealized reforms most needed by Brazil's poor, including more equitable land distribution. The

Aliança Nacional Libertadora (National Liberation Alliance) staged a revolt in 1935, after which Vargas outlawed it. This launched years of political repression, during which many leftist intellectuals were persecuted. Prior to the planned presidential elections of 1937, Vargas assumed dictatorial powers, claiming that this was necessary for political stability. His authoritarian Estado Novo (New State), which continued until 1945, was supported by many who believed that imposing social order was paramount. During this period the government censored the press and imprisoned its critics. Vargas's swing to the political right tempered much of the progressive energy that marked the early years of his regime.

A conservative political party gained broad influence in Brazil during the 1930s, as a counterweight to the left-wing Aliança Nacional Libertadora. Ação Integralista Brasileira (AIB; Brazilian Integralist Action) was a Catholic anticommunist party modeled on European fascist movements. Among its outspoken members was Belisário Penna, the leader of Brazil's rural public health reforms during the 1920s and Vargas's health minister from 1930 to 1933. Penna characterized the AIB as aiming to "liberate the nation from sordid Bolshevik materialism and create a new patrician mentality [based on] moral Christian principles, the security and morality of families, and the unity and integrity of the nation."[10] He blamed national disunity, fostered by the republic's decentralized government, for the advance of communism in Brazil. To explain how his membership in the AIB meshed with his earlier progressive goals, Penna argued that fascism would promote genuine national development addressing the needs of all Brazilians. Vargas distanced himself from some aspects of the *integralista* platform. Yet his Estado Novo dictatorship conceded that authoritarian rule was necessary for the state to pursue rational solutions to the nation's problems.

The fluctuating political climate under Vargas is evident in literary and scholarly writing about the Nordeste region in the 1930s. Vargas's revolution coincided with a flowering of important novels that portrayed the bleak poverty of many Nordestinos' lives and the need for social reform. Some of these focused specifically on droughts and the migration of sertanejos from their homelands. *O Quinze*, published in 1930, examines the hardships suffered by sertanejo migrants from the municipality of Quixadá, Ceará, during the 1915 drought. *Cearense* author Raquel de Queiroz had been raised in the sertão, and drought was a formative experience in her childhood. Eight years later, Graciliano Ramos, who spent his childhood in the *sertões* of Alagoas and Pernambuco, produced another significant "drought novel." *Vidas Secas* (Parched lives) employs spare prose to describe sertanejos adapted to life in a harsh climate. Ramos's humble protagonists are ridiculed by people who view them as unsophisticated. Their energies are absorbed by the need to survive, which seems to drain them of higher ambitions. By the time *Vidas Secas* was published, the reformist fervor of the early 1930s had been replaced by the repressive Estado Novo regime, and Ramos's unmistakable criticism of social conditions in the sertão led to his imprisonment. During that period, several authors whose work criticized social conditions in rural Brazil were persecuted to varying degrees; these included Gilberto Freyre, whose book *Nordeste* highlighted the environmental destruction and malnutrition that accompanied sugarcane monoculture, and Ruy Coutinho, labeled a communist for his biting critique of food scarcity and malnutrition in Brazil.[11]

· · · · · ·

To spur Brazil's economic recovery during the global recession of the early 1930s, Vargas adopted a policy of import-substitution industrialization. This directed the majority of his administration's resources to urban areas, which pleased his political base. But in the minds of some contemporaries and several historians, Vargas's industrialization agenda jeopardized rural Brazil. José Américo Almeida (the author of *A Bagaceira*) believed that Vargas's legislation to improve the conditions of urban workers "endangered the countryside, because it favored urban laborers so much and initiated the exodus from the countryside to the city."[12] Historian Robert Levine echoes this assessment, observing that Vargas's labor policies had little effect in rural areas, where oligarchs routinely disregarded the law. Vargas's focus on urban workers increased the gap in economic opportunity between urban and rural regions, Levine finds, precipitating increased migration from the backlands to the south's major cities.[13]

Nevertheless, Vargas's tours of the Nordeste in 1933 and 1940 showed him to have a devoted following there.[14] The president first visited the region following a severe two-year drought, and he made a public commitment to continue funding IFOCS at a high level in order to execute a coherent plan of drought works. He argued that northeastern drought works had to be a national priority even during economically difficult times because Brazil could not afford to continue losing workers to starvation and disease. In order to advance as an industrial nation, the sparsely populated country needed the productive capacity of all its citizens.[15] Such pronouncements contributed to the popular perception of Vargas as a "protector" of drought victims. Archives of ordinary Brazilians' correspondence with the president during the 1930s indicate the hope placed in his leadership by many individuals and communities in the sertão. Aspiring drought migrants asked Vargas for free passage on the national rail line, Lloyd Brasileiro, so that they could find work to support their starving families. Others beseeched him to supply return passage, having been tricked into exploitative labor arrangements in the south and lacking funds to return home. The vast majority of these requests were denied, though Vargas's respondents expressed considerable sympathy for the petitioners' plight.[16]

Vargas received many requests to authorize the opening of new reservoir and road projects. Some writers contacted the president for mediation when they felt that IFOCS exhibited political favoritism in its acceptance or rejection of drought works proposals. Correspondents also pleaded that Vargas order the reemployment of devoted fathers who had been dismissed from IFOCS as office staff or construction personnel. Desperate communities begged him to ensure that they received market-rate indemnification from IFOCS for homes and farmland about to be inundated by reservoirs. The indemnification process was often delayed due to residents' lack of legal title to the land where they lived and farmed. All such queries were forwarded from the presidential palace to the drought agency, and many received a detailed response, but these communications rarely resulted in a change of plans or provision of aid. Drought Inspector Luiz Vieira and his staff explained patiently and repeatedly to Vargas's many petitioners the policies, priorities, and funding limitations that compelled the drought agency to select one project over another. In some cases, petitioners were told that they might qualify for assistance under the agency's cooperative construction program, in which costs were shared by the government and landowners, and that they should apply for that.[17]

Despite the widespread public perception of Vargas as patriotic "father" of all sertanejos, he appears to have acted pragmatically in the sertão. He devoted personal attention and federal funds to it when droughts highlighted the precarious circumstances of most sertanejos, but directed resources to more influential parts of the country—particularly the south's industrializing cities—when conditions in the northeast returned to the status quo. As one example, the 1934 constitution, which aimed to codify the priorities of Vargas's revolutionary movement, allocated 4 percent of federal tax revenues "permanently" for drought aid, but this provision was revoked in the 1937 constitution drafted under Vargas's dictatorial Estado Novo. The moderate reformism of the Vargas years is reflected in IFOCS's achievements during the 1930s. The drought agency began hiring agronomists in 1932, during a devastating drought. By the mid-1940s, the agency had made modest headway in providing landowners with irrigation, but little change had been made to the landholding structure that left many sertanejo farm workers vulnerable to drought. This approach was consistent with Vargas's emphasis on rational solutions to social problems, which aimed to limit dissent and social rupture. IFOCS's regional development plans continued to accommodate the interests of estate owners, even though those often ran counter to the needs of smallholders and landless farmers whom its staff also intended to serve.

Assessments of Vargas's legacy in the sertão run the gamut from unmitigated praise to guarded criticism. Engineer Rui de Lima e Silva, impressed by IFOCS's progress during the early 1930s, wrote that the agency had exceeded all expectations thanks to the hard work and self-sacrifice of Brazilian engineers willing to live for a time in the sertão. "Suffice it to say that essentially all of the projects organized by the North American mission were abandoned as inappropriate or unexecutable," he wrote, referring disparagingly to the work of American engineering firms hired by the federal drought agency during the early 1920s.[18]

The following *cordel* (folk poem) written by a prominent Bahian poet is typical in its praise of Getúlio's efforts on behalf of needy Nordestinos:

| | |
|---|---|
| *Protegeu orfãos, viuvas,* | He protected orphans, widows, |
| *Famílias desamparadas* | Abandoned families, |
| *Deu Abono de família* | Gave benefits to families |
| *Melhorou as classes armadas* | And improved the armed forces |
| *Beneficou o nordeste* | Benefited the northeast |
| *Combateu a sêca e a peste* | Combated drought and disease |
| *Em todas as suas camadas.* | At every level. |
| | |
| *Do Palácio do Catete* | From the Palace at Catete [in Rio] |
| *Viajou um certo dia* | He traveled one day |
| *Percorreu o norte inteiro* | Traversing the entire north |
| *Para ver o que havia* | To see what was happening. |
| *Cada logar que passava* | In every place that he passed |
| *De tudo que precisava* | He was happy to do |
| *Ele contente fazia.* | Whatever was needed.[19] |

The poem ends by calling Vargas's critics ambitious traitors, saying that all Brazilians benefited from his governance. Notably, in his unpublished memoir the poet explains that he was not personally an enthusiastic fan of Vargas but that poems praising the president sold well because many Nordestinos admired him.[20]

## José Américo Almeida, Regional Patriot

In November 1930 Vargas appointed energetic Nordestino reformer José Américo Almeida as minístro de viação e obras públicas (MVOP; minister of transportation and public works), a role that included oversight of the drought agency along with other responsibilities. The new minister had grown up in the highland sertão of Paraíba state, in an established family. He graduated from Recife's law school in 1908, then returned to Paraíba and became involved in state politics. Almeida was fiercely loyal to the Pessoa family, one of Paraíba's reigning political clans, and these ties were one reason for his selection as minister.[21] Vargas's vice presidential running mate, Paraíban governor João Pessoa, had been assassinated while campaigning. The president portrayed his nomination of Almeida to the ministerial post as compensation to the northeast—and Paraíba in particular—for sacrificing a native son to Vargas's political ambitions. Prior to his appointment to the ministerial post, Almeida had acted as chief of Vargas's revolutionary movement for the north and northeast, then as *interventor* (unelected governor, appointed by Vargas) for Paraíba, and finally as chief of the Governo Central do Norte (Central Government of the North) within Vargas's revolutionary administration.[22]

Almeida was well known nationally before 1930, due primarily to the publication of his 1928 novel *A Bagaceira*. Earlier he had written a book in praise of Epitácio Pessoa's drought alleviation efforts, called *A Paraíba e Seus Problemas* (Paraíba and its problems). He published it in 1923 to draw attention to President Artur Bernardes's severe reduction in funding for IFOCS, which ultimately reversed the progress that Pessoa had made in the sertão. Almeida marshaled testimony from many people familiar with Pessoa's drought alleviation efforts to refute the common accusation that IFOCS's early 1920s projects were disastrous and rife with corruption. His sources emphasized the potential of drought works to dramatically improve the northeast's economic and social health.[23] *A Paraíba e Seus Problemas* underscored the value of national investment in the sertão and the folly of neglecting such a rich and expansive region. Almeida enumerated his state's favorable geographic features as an argument against its "abandonment" by the federal government, a word chosen to evoke "the destructiveness of colonial politics, the excessive centralization during the monarchy, and the discredited [state] autonomy of the republican era, all obstacles to our kind of progress. The data confirming this negligence are interesting and highlight to a great extent the lack of assistance from our leaders."[24] He depicted sertanejos as honorable and hardworking, held back only by their sometimes inhospitable environment. He argued that to combat the national government's tendency to neglect the northeast, more Nordestinos like Epitácio Pessoa should ascend to the presidency.

In 1931, as the new MVOP, Almeida laid out his priorities for the drought works agency. His main goal was to execute Pessoa's plan for reservoir development, since none of Pessoa's major reservoirs had been completed, many works begun during his administration had been washed out by floods, and the remainder had deteriorated from neglect. Almeida wanted IFOCS to focus on reservoirs and irrigation, with roads as a secondary emphasis to provide access to construction sites and help move drought-affected populations to fertile areas. He believed irrigation was the only way to significantly improve sertanejos' ability to withstand droughts, and he argued that all states in the drought zone had areas with as much productive potential as São Paulo. There was no need for sertanejos to migrate as long as those subregions could be cultivated.[25]

Almeida also thought that stocking reservoirs with fish should become a standard component of IFOCS's work. This would foster an economically and nutritionally viable alternative to the ranching industry, which suffered substantial losses during droughts. The minister viewed fish as a more efficient use of the sertão's resources than cattle. One hectare of reservoir water could support two thousand kilograms of fish flesh, he claimed, whereas one hectare of pasture yielded at most one hundred kilograms of beef.[26] The average sertanejo's diet was deficient in protein, which he obtained from poor-quality dried beef and beans. Like Vargas, Almeida believed that improving the diet of agricultural laborers would increase their productivity and thereby aid regional industrialization, by providing a more reliable food supply for urban workers. From 1932 to 1943, IFOCS stocked its reservoirs with over 550,000 fish of fourteen species.[27]

In April 1931 Almeida hired Artur Fragoso de Lima Campos to direct IFOCS, but Lima Campos only remained in the position for one year. Civil engineer Luiz Vieira took up the post in April 1932 and remained in it for a decade. (He resumed the position in 1946–47.) Vieira hailed from Rio de Janeiro and had graduated from the Escola Politécnica. He would spend his entire career with the drought agency. Under Vieira's guidance, IFOCS pursued a dual strategy of building large public reservoirs and partially funding smaller ones on private property; these were intended to keep as many agricultural workers as possible from leaving their home regions during droughts. Almeida and Vieira also established an irrigation plan focusing on four watershed basins in the sertão: Acarahú and Jaguaribe in Ceará; Alto-Piranhas in Paraíba, and Baixo-Assu in Rio Grande do Norte. In 1935 Vieira sponsored a survey of part of the São Francisco River valley; surveyors used air photography to assess the river's potential for irrigation, navigation, and energy production.

During the devastating drought of 1931–32, Almeida traveled from his Rio de Janeiro office to Ceará to help organize aid efforts. The scourge was unusual in its geographic extent, affecting over 650,000 square kilometers. This made moving sertanejos to less affected areas, as had been attempted during previous crises, very difficult. The challenge of aiding drought victims was exacerbated by a substantial population increase in the sertão during the early twentieth century, the result (in part) of an extended period with few severe droughts. By the mid-1930s the sertanejo population was estimated to be 2,636,500, with the majority concentrated in river valleys.[28] In order to prevent a mass exodus, the Ministry of Transportation and Public Works established *campos de assistência* (refugee camps) where aid was distributed—guarded areas referred to by their reluctant inhabitants as "corrals." Two of these, near Crato and Fortaleza, Ceará, held over 100,000 people each. Survivors of the camps recall the foul stench, putrefying food, and rampant disease as well as the constant presence of death. Mass graves were shallow and often plagued by roving dogs and vultures.[29]

At the camps, sertanejos were vaccinated against typhoid, dysentery, and smallpox to halt epidemic outbreaks, a policy that sometimes required coercion. One physician forged a letter from the popular religious leader Padre Cícero saying that vaccinations were good. He showed this to a few influential men in the camps, and they persuaded their followers to submit to the vaccine.[30] Almeida had seeds and agricultural tools distributed to drought refugees in hopes that they could farm productively once the crisis abated. He gave relief money to bishops for distribution in remote areas where the Catholic Church was the only institutional presence, a practice that ran counter to Brazil's laws for federal aid distribution and for which Almeida was criticized.[31] Despite these significant efforts, tens of thousands of people were believed to have died as a result of the 1931–32 drought. The works completed prior to 1930 had been insufficient to spare them.

Because he visited the drought zone personally and seemed genuinely concerned about the plight of drought migrants, Almeida earned a reputation among sertanejos as one of Brazil's most honest and hardworking politicians. Vieira credited his boss with being the first official in Brazil's national government to give drought aid the resources it deserved;[32] speaking to Rio de Janeiro's Clube de Engenharia in 1935, Vieira described Almeida as a devoted son of the northeast, trying to "halt the exodus of the masses, their annihilation by hunger, or the invasion of cities by the desperate multitude, along with the dangers of fatal epidemics, banditry, and all the other miseries witnessed during prior episodes."[33] Vieira himself moved IFOCS's headquarters to Fortaleza, Ceará, in order to oversee the agency's expanding operations and provide moral support to staff working in the stricken region. Many commentators at the time saw departing the comforts of Rio de Janeiro for the health risks of the suffering Northeast as heroic.

Almeida and Vieira opened road and reservoir projects to employ drought refugees, which they termed *frentes de trabalho* (work fronts). These served to remove the migrants from cities, where they were seen as a threat to residents' health and security, and temporarily increased IFOCS's number of personnel dramatically. Prior to the 1931–32 drought, IFOCS had a staff of ninety-one: forty-five in its Rio de Janeiro headquarters; twenty-two in Fortaleza, Ceará; fifteen in João Pessoa, Paraíba; and nine in Salvador, Bahia. Each of the four offices was directed by an engineer, as required by law.[34] In March 1932, the agency's total personnel had expanded to seven thousand; by the end of that year, IFOCS employed 220,000 people. The men temporarily "enlisted" on the federally financed projects were paid in basic foods (rice, beans, and the cakes of brown sugar called *rapadura*) to sustain their families. Ninety-three engineers and fifty-three technical assistants (men who had fewer educational qualifications than engineers, but substantial practical experience) directed its operations, overseeing hundreds of manual laborers each. If enough starving men came to them in search of work, some drought agency personnel agreed to form *frentes* prior to receiving authorization from their central offices. The staff member sought immediate funds from a private *fornecedor*, and hoped to receive reimbursement eventually from the federal government.[35] Almeida estimated that IFOCS's 1932 workforce enabled nearly 900,000 drought victims to obtain food rations, assuming three dependents per worker.

Conditions around worksite encampments were grim. Everything was scarce—but particularly water, which sometimes had to be transported ten kilometers over poor roadways in sufficient quantity to sustain thousands of people. The camps were rife with disease due to the lack of clean water, a situation made worse by overcrowding and sertanejos' physical weakness from hunger. These conditions led to outbreaks of typhoid and dysentery, which a special public health commission sent from Rio de Janeiro managed to curb with the aid of vaccines and sanitary education. Still, Vieira calculated that 15,909 people died while residing in IFOCS's refugee camps in 1932–33; almost two-thirds of them were children.[36] When tools were unavailable to employ the number of families that arrived at a worksite, Vieira instructed his staff to provide charitable rations rather than turn people away, though he deemed such aid humiliating to men who valued self-sufficiency. Families without male heads of household posed a particular challenge for the agency. In contrast to some of his predecessors, Vieira petitioned his superiors to allow women or teenage boys from such families to enroll as workers, arguing that the likely alternative was the "moral degradation" of girls who would turn to prostitution to support their starving mother and siblings. By November 1932, 15 percent of IFOCS's workforce comprised boys between the ages of ten and fifteen, often orphans in charge of younger children.[37]

Vieira described the goals of IFOCS's drought assistance programs in different ways depending on his audience. At times he depicted the launching of new projects during the 1931–32 drought as a form of charity undertaken with almost too few resources to succeed. At other times he explained that the rationale for providing drought refugees with paid work was to prevent their migration from the sertão. The mass exodus destabilized areas where the refugees settled, and it left the drought zone with insufficient labor once fertile years returned. When discussing IFOCS's early 1930s expansion with fellow engineers, Vieira acknowledged that he had tried to take advantage of the cheap labor and increased funding made available by the drought to accomplish as much construction as possible.

At the height of the 1931–32 drought Vargas allowed IFOCS to spend 170,000 contos. This was seventeen times what the agency's budget had been when he assumed executive power. Vieira spent as much from 1931 to 1934 as his predecessors had spent from 1909 to 1930 (although almost 75 percent of those earlier expenditures occurred during Pessoa's presidency, 1920–22). From 1931 to 1934, IFOCS increased the capacity of public reservoirs in the northeast by more than twice what it had been up to 1930. Vieira bragged that his construction teams completed in only a few months works that British and American engineers had abandoned in the 1920s (presumably due to lack of funds once Bernardes's administration began).[38] Reflecting in mid-1934 on IFOCS's progress under his direction, Almeida emphasized the breadth and speed of projects undertaken and completed by the agency.

Nonetheless, IFOCS's accomplishments during and immediately after the 1931–32 drought indicate the substantial preference that its managing engineers continued to give dams over agricultural projects. From 1931 to 1935, Vieira's staff oversaw the completion of twenty-nine public reservoirs with a capacity of almost 1.3 billion cubic meters of water. They also helped to construct forty-nine reservoirs on private land (mainly in Ceará) holding fifty-nine million cubic meters of water. Thus, by the end of Vieira's first term, IFOCS had brought the total capacity of all of its reservoirs (public and private) to three billion cubic meters, or 20 percent of the agency's total goal for water storage; this was five times what the public reservoir capacity had been in 1930.[39] Vieira hoped to increase the holding capacity of public reservoirs to fifteen billion cubic meters, though this was not realized for many decades. (In the early 1990s, the northeast had fourteen billion cubic meters of water stored, mainly in Ceará.[40])

Vieira estimated the total irrigable land area within the drought zone to be at least 300,000 hectares (over 740,000 acres)—with more arable land along the São Francisco River, provided that an economical means of pumping that water could be developed. Following the 1931–32 drought, he claimed that irrigation works were underway for 12,900 hectares (roughly 31,900 acres) near IFOCS's four agricultural posts. However, by 1940 only five thousand hectares (12,350 acres) had actually been irrigated. By 1943, the agency's networks of irrigation canals covered 354 kilometers, supplying ten thousand hectares of farm land (24,700 acres). IFOCS had doubled its irrigation network in a mere three years, yet this supplied water to only 3 percent of the sertão's estimated irrigable area.[41] Engineers' and landowners' priorities continued to govern the drought agency, favoring reservoir and road construction over expanded irrigation.

· · · · · ·

Despite Almeida's enthusiasm for agricultural reform, IFOCS made only token progress toward irrigating the sertão in the 1930s. Although Vargas made a great show of attending to sertanejos during the drought, he was not willing to maintain that level of support for drought works once the crisis abated. By the mid-1930s the politically peripheral sertão had lost its place in the national spotlight, and development efforts slowed. In 1934 the inspectorate's budget fell to 47,000 contos—significantly higher than before the Vargas administration, but a far cry from the amount apportioned during the crisis.[42] Almeida felt that Vargas was not upholding the ideals of the 1930 revolution, which included promoting economic independence and progress for all Brazilians.

Almeida resigned from his post as MVOP in 1934. He had become disillusioned with Vargas, whose dedication to aiding rural Nordestinos proved fickle. In 1937 Almeida ran for president against Vargas, proclaiming that it was time for Brazil to address the needs of the poor, since the wealthy were already well taken care of. For this he was seen by some as quasi-communist, and Vargas encouraged this image since it helped to justify his declaration of authoritarian rule (on the grounds that communism had become a threat to the nation). Following his resignation from the ministry, Almeida continued his political career in other capacities and remained committed to responsive democratic government. In 1945 he gave an interview to the Rio de Janeiro newspaper *Correio da Manhã* that criticized Vargas's censorship of the Brazilian press; the publication of the interview stimulated more open criticism, helping to bring Vargas's eight-year dictatorship to a close. In 1947 Almeida became a senator for Paraíba and subsequently served as its governor. In that capacity, he instituted policies to reduce nepotism in state appointments. When a drought began in 1953, Vargas invited Almeida to return as MVOP, aiming to emphasize his genuine concern for those affected by the calamity. The highly regarded Nordestino accepted out of a desire to help victims of the drought and a lingering sense of duty to Vargas for their prior collaborations.

## Struggles for Authority over IFOCS: Agronomists versus Engineers

Even though IFOCS's projects were disproportionately weighted toward engineering works, the agency's engineers felt that they were undervalued by Vargas's administration. In 1937, after hearing Ildephonso Simões Lopes deliver a speech about Brazil's drought problem, representatives of the agency's engineering staff asked him to exercise his influence with Vargas. Simões Lopes was a prominent political figure who had trained as a civil engineer; he had been minister of agriculture, industry, and commerce under President Pessôa and served as one of the three commissioners sent to review the drought agency's progress in 1922. Under Vargas, he became director of the Bank of Brazil. IFOCS's engineers appealed to him as someone who understood the importance of their work in defense of Nordestinos, through which they claimed to combat inflation, social disorder, the "dishonoring of virgins," and the "vile exploitation of those in misery," among other ills.[43] They wanted their compensation brought into line with that of civil engineers in other inspectorates overseen by the Ministry of Transportation and Public Works, dealing with national roads, ports, and rivers.

Yet when Vargas first came to power, as president of Brazil's National Agriculture Society (1926–31) Simões Lopes had warned the president that improved agriculture was more important for the sertão than expanded engineering works. He cautioned that "without a doubt, [lack of water] is the ultimate cause [of droughts in the northeast], but the immediate sources of damage turn out to be not the lack of water but . . . a lack of subsistence and forage crops." Hunger and weakness made people and cattle highly vulnerable to drought conditions long before thirst set in. A native of the ranching state of Rio Grande do Sul, Simões Lopes was particularly concerned about inefficient practices of supplying forage to cattle. He recommended outreach by agronomists to improve forage crops, particularly in years when a drought threatened. "The solution to the problem is a question of practical agriculture and not of engineering works," he asserted.[44]

Juarez Tavora, minister of agriculture from 1932 to 1934 and a former general in Vargas's revolutionary Aliança Liberal (Liberal Alliance) in the northeast, also contended that IFOCS's staff possessed insufficient agronomic expertise. Like Almeida, Tavora hailed from Paraíba. He had been Vargas's first choice as MVOP but turned that position down to remain involved with military affairs. When Tavora's relationship with subordinate officers soured, Almeida nominated him to become minister of agriculture.[45] In that role Tavora wrote a brief report on the "Necessity of collaboration by agronomists in the drainage and irrigation works in progress in the Northeast," citing various problems that arose as a result of having civil engineers oversee irrigation projects. These included constructing gigantic reservoirs with little irrigable land alongside them, while overlooking easily irrigable areas when planning reservoir construction, and laying canals in places where fragile soils could not tolerate irrigation.[46] During the 1931–32 drought, Tavora recommended that IFOCS increase its support for irrigated cultivation in the sertão and establish farming colonies outside of the drought zone.[47]

In response, Almeida solicited reports from agronomists about how to establish colonies where drought refugees could settle and learn modern farming techniques. Following five months of travel to encampments across the northeast (and a lengthy bout with dysentery), Evaristo Leitão provided his recommendations. Underutilized land suitable for at least one hundred families, with the possibility of expansion, should be surveyed and divided into lots of twenty to thirty hectares. Each family would be given a modest, hygienic home of concrete, brick, and tile and access to bank credit. The settlements would include crop storage facilities, a small store, a pharmacy, a health clinic and ambulance, an agricultural experiment station, a meteo- rologic post, and a school adapted to the needs of rural students, to improve their future productivity as farmers. Other construction works would pro- vide roads, sanitation infrastructure, and factories to process agricultural products like cotton. The government should offer technical instruction to start the settlers on the road to economic independence. Initial funds for resettlement projects should come from a tax on agricultural exports. Taxes collected on the products sold by the colony would repay the state's establishment and ongoing administration of each community. Leitão's detailed description encapsulates agronomists' optimistic view of what ir- rigated smallholding could achieve in the sertão to combat the range of ills plaguing rural workers—from illiteracy to disease and poverty. "In a few years," he proclaimed, "we will have combatted nomadism and integrated into the national community this valorous contingent of citizens who, cur- rently and for many years, have comprised one of the gravest problems for our country."[48] Under Almeida's administration, only a small number of drought refugees were settled in agricultural colonies of this type outside the drought zone. When he returned as MVOP during the 1953 drought, Almeida remarked that the considerable delays surrounding irrigated colo- nization projects indicated the strong opposition to them among northeast elites, who perceived smallholder independence as a threat to their system of tenant labor.[49]

Only in the late 1930s did drought inspector Vieira begin to emphasize irrigation as a central element in IFOCS's drought alleviation strategy. A 1937 law (no. 508) required irrigation canals to be initiated simultaneously with the construction of all new reservoir projects. The law made coopera- tive government funds available to farmers who installed pumps to irrigate at least five hectares of land.[50] In his public presentations, Vieira began to portray sertão development as a two-pronged endeavor. Reservoirs were the first defense against droughts; along with their most important purpose—

namely, water storage—they could be used to regulate river flow, generate energy, and stock fish. Vieira described irrigation as a second line of defense, to overcome food scarcity and avoid the famines and mass migration that historically accompanied severe droughts.[51]

Engineers' dominance of IFOCS almost ended in 1942, when the minister of agriculture tried to gain control of the agency. Vargas had created the Ministry of Agriculture in 1930 as part of an extensive reorganization of the federal government. Previously agriculture had been overseen by ministries that also had other responsibilities—such as the Ministry of Agriculture, Industry and Commerce, which existed from 1890 to 1930.[52] In 1933 the agricultural ministry became responsible for regulating the new agronomic profession and established the first national standards for the practice of agronomy. These dealt primarily with which government jobs needed to be filled by trained agronomists; among these were oversight of agricultural projects requiring small-scale dam and road building—projects that in other circumstances would be directed by civil engineers.

Because of the drought agency's growing commitment to irrigation, Agriculture Minister Apolônio Sales proposed in 1942 that IFOCS be transferred to his domain. (Sales had been an advocate of irrigated cultivation since his employment at the Usina Catende sugar mill in Pernambuco.) The rationale for such a reorganization had been discussed by high-ranking officials of several ministries for over a year.[53] Vargas's Departamento de Administração do Serviço Público (DASP; Department of Administration of Public Services), responsible for rationalizing federal bureaucracy, supported Sales's proposal.[54] According to DASP, which had been established following training received by Brazilian bureaucrats in Washington, DC, IFOCS's engineering works had been constructed to improve northeastern agriculture. Since most dams were complete by the early 1940s, the primary remaining task was to construct irrigation canals. Based on the priorities established for IFOCS by Almeida in 1931, many of its activities—such as geographic surveys, the establishment of botanic gardens and agricultural posts, and research into fish cultivation—fell more appropriately under the domain of the Ministry of Agriculture. That ministry already included a National Department of Vegetable and Animal Production, a forest service, and a meteorological service, all of which could be productively employed in tackling the northeast's drought problem. As the DASP director noted, "The transition of IFOCS into the Ministry of Agriculture would complete [Vargas's] reorganization plan, which obeys the principles of uniformity, convergence of forces, and identity of goals."[55]

DASP suggested renaming IFOCS the Serviço de Obras Contra as Secas (Service for Works to Combat Droughts) because the designation "service" indicated a broader range of responsibilities than "inspectorate." An alternate renaming that DASP proposed was the Departamento Nacional de Recuperação de Solos (National Department for Recuperation of the Soil), since combating soil erosion was one of IFOCS's most pressing tasks. To compensate the Ministry of Transportation and Public Works for losing IFOCS, DASP suggested that it absorb the National Department of Mineral Production from the Ministry of Agriculture.

Despite DASP's support of Sales's proposal, the Ministry of Transportation and Public Works managed to retain control over the drought agency and successfully defined IFOCS's purpose as "the realization of all works intended to prevent and attenuate the effects of droughts in the North and Northeast," stressing that the agency had a broader mandate than merely the promotion of improved agricultural production.[56] As a result of this debate, IFOCS was renamed the Departamento Nacional de Obras Contra as Secas (DNOCS; National Department for Works to Combat Droughts) in 1945, giving it higher status within the MVOP. DNOCS was divided into four districts, headquartered in Arcoverde, Alagoas; Fortaleza, Ceará; João Pessoa, Paraíba; and Salvador, Bahia. At the same time, its agricultural service, previously called the Commission of Services Complementary to IFOCS, was renamed the Agro-Industrial Service to emphasize its primary focus.

The debate over which ministry should oversee drought works helps to explain why agronomists remained relatively ineffective within IFOCS throughout the 1930s despite the interest in agricultural development expressed by Almeida and others. The ministry that retained control over the drought agency was more committed to and expert in civil engineering than agricultural endeavors. Its professional priorities conveniently accommodated the interests of influential politicians in the region: sustaining cattle while maintaining tenant labor on extensive estates. Agronomist Paulo de Brito Guerra, who directed the research institute at the São Gonçalo agricultural post during its early decades (and became the first agronomist to oversee a DNOCS division that included engineers), described the challenges that insufficient funding posed to IFOCS's agricultural service: agronomists went for months without pay or electricity and often had to furnish their own transportation to field sites; they relied on manual labor until the 1980s, whereas drought agency engineers used animal and machine power for road construction beginning in the late 1930s.[57] Because the agricultural service often could not afford to hire a sufficiently large workforce, some of the land on its extension posts was rented to farmers rather than used for research.[58]

One reason DNOCS was reluctant to fund research was that it took many years and might not yield tangible benefits (although Guerra asserted that U.S. investment in improved agricultural production eventually reaped tremendous economic rewards). Particularly during the global export slump of the 1930s, long-term research goals were hard to justify in many Latin American countries.[59] But agricultural extension along with irrigation works and the surveys that preceded them also suffered because they were not valued within the northeast's system of political patronage. In Guerra's words, the agricultural service's projects were "invisible and unsuitable for inauguration ceremonies."[60] Still, the service's surveyors recalled meeting little resistance from small farmers and estate owners after the initial phase of their operations, when it became clear to these clients that irrigated agriculture could be profitable.[61]

## Development and Disillusion

Almeida believed irrigation should become a core focus of IFOCS's work, yet during his brief term as minister of transportation and public works he did not manage to shift its priorities substantially from dam building. The agricultural service established within the agency in 1932 had only a modest impact on the sertão's physical and social landscape during its first decade; IFOCS's managing engineers awarded the service less than 3 percent of the annual budget. According to Guerra, half of the service's budget was spent by the central agricultural post, São Gonçalo, which left very little funding for irrigation canals and agricultural extension elsewhere—and even São Gonçalo's funding was "laughable."[62] IFOCS remained dominated by civil engineers whose capabilities and professional allegiances led them to pursue reservoir and road construction above all other efforts to mitigate the drought problem. Engineers were chosen for administrative positions partly because their training at positivist mining or military schools was thought to make them good managers. Their profession was also held in higher regard in 1930s Brazil than was the new discipline of agronomy.

Ultimately, irrigation received insufficient support from the drought agency during its first decades because it was threatening to the landholding elite. Smallholder irrigated colonies aimed to provide farmers with a secure means of supporting themselves, which would reduce their dependence on the sertão's power brokers. Yet it was this very dependence that made *coronéis* powerful within local systems of patronage. For DNOCS to substantially expand smallholding in the sertão, it would have had to expropriate and redistribute extensive land areas around reservoirs. Any effort in this direction would have harmed the agency's relationship with many regional politicians. Since DNOCS's managing engineers were not particularly invested in irrigation themselves, they focused on the projects that advanced both their own professional agendas and the interests of ranchers and export farmers whose support their agency required for its own continuation. The benefits resulting from IFOCS's construction of reservoirs and (on a smaller scale) irrigation canals accrued disproportionately to large landowners, who saw their productivity rise and their land values increase. Perhaps inadvertently, Vargas's drought bureaucracy managed to strengthen the existing social order.

# 5  Watering Brazil's Desert
## Agronomists and Sertão Reform, 1932–1955

In 1940, national newspapers effusively praised the completion of irrigation canals around the Forquilha Reservoir in Ceará. The *Diário Carioca* proclaimed that drought agency technocrats were revitalizing and civilizing the sertanejo "race," which had felt abandoned by fellow Brazilians and punished by God—a perception that led to a loss of moral sentiment. According to the newspaper's editorialist, "Irrigation water represents a rebirth and a modern organization of regional life in every aspect"—as exemplified by the laudable impact of U.S. agricultural experiment stations run by a variety of government agencies in that forward-thinking republic. The "fertile but ungrateful" sertão required much more to rehabilitate it, but irrigation networks were an essential and admirable first step. Smallholding would bring sertanejo sharecroppers out of their rustic "backwardness" and form the nucleus of well-appointed towns, complete with electricity, parks, a cinema, and other modern amenities.[1]

Agronomists who joined the drought agency in the 1930s emphasized the centrality of irrigation networks to any reorganization of the sertão's economy and society. They soon discovered, however, that even the sertão's most precarious households were reluctant to adopt the rigors of irrigated cultivation except during drought crises, when no other means of producing food for their families and livestock were available. Thus, agronomists came to view regional development as fundamentally a process of acculturation. Sertanejo farmers must be persuaded to alter their work regimes to protect their families from the unpredictable threat of drought and starvation, rather than following the culturally prescribed "law of least work" that, some believed, characterized sertanejo society.[2] More formidably, large landowners had to be persuaded of their moral and patriotic duty to ensure the survival of their tenant workers' families when droughts struck. Archival records of irrigation projects illuminate the "middle politics" of agronomists from the Departamento Nacional de Obras Contra as Secas (DNOCS; National Department for Works to Combat Droughts) who supervised agricultural posts and irrigated settlements in the 1940s and 1950s.

While defending poor farmers against the most exploitative labor and landholding practices, they also confronted many farmers' disregard for their recommendations about how irrigated land should be managed in the interest of efficient production.

Agronomists worked directly with sertanejos on agricultural extension posts, which made them more aware of the socioeconomic dimensions of drought suffering than DNOCS's managing engineers were. The first two directors of the agency's agricultural service, José Augusto Trinidade and José Guimarães Duque, expressed sympathy for the plight of small farmers buffeted by drought, and both viewed irrigated smallholding as the best mode of production for ordinary sertanejos' welfare and the region's economy. Yet during the 1930s and 1940s, Trinidade and Duque rarely emphasized the predominance of large estates in the sertão as a contributor to the human suffering precipitated by harvest failure. Instead they blamed the recurrent tragedy primarily on sertanejos' indiscipline. Emphasizing the need to acculturate farmers to more intensive, cooperative farming methods suited the authoritarian paternalism of President Getúlio Vargas's Estado Novo regime (1937–45).

Trinidade explained the lack of irrigated agriculture in the sertão as the result of cultural factors shaped by the environment. He believed that the predominance of years with abundant rainfall made sertanejo farmers reluctant to invest in irrigation technologies that were only necessary during droughts. His focus on improved technical education implied that farmers' inability to withstand droughts was largely the result of their own ignorance and poor planning rather than of insufficient control over land and water. Duque carried Trinidade's cultural explanation for drought crises a step further, arguing that sertanejos lacked the cooperative spirit necessary for successful irrigation. He also thought that making small farmers more receptive to irrigation, through educational outreach, would enable them to avoid the food shortages that normally accompanied droughts. By the 1950s, under the ambitious modernizing administration of Juscelino Kubitschek, Duque had begun calling more stridently for land redistribution as essential to the broader social transformation that irrigated smallholding might engender.

## José Augusto Trinidade and José Guimarães Duque at the CSC

In response to the 1932 drought, José Américo Almeida had created the Comissão Técnica de Reflorestamento e Postos Agrícolas do Nordeste

(Northeast Technical Commission for Reforestation and Agricultural Posts) within the Inspetaria Federal de Obras Contra as Secas (IFOCS; Federal Inspectorate for Works to Combat Droughts) and named José Augusto Trinidade as its first director. A native of Minas Gerais state whose wife was from "one of [Paraíba's] most traditional families,"[3] Trinidade had studied at the agronomy school in Pinheiro, Rio de Janeiro, then joined the new Ministry of Agriculture and taught briefly in Minas Gerais's agronomy school at Viçosa. He identified four main tasks for IFOCS's agricultural commission: it would sponsor studies of the sertão's native plants and soils with an eye to improving agricultural productivity; it would promote and support irrigated agriculture and provide extension services to as many sertanejo farmers as possible; it would research methods of preserving cattle forage; and it would introduce mechanized agriculture.

Trinidade hired three recent graduates of the Escola Nacional de Agronomia to study the sertão's soils. He planted fields of commercially useful palms (mainly oiticica, from which a nut oil was harvested for use in manufacturing dyes) and started tree nurseries on the banks of reservoirs. In 1933 he hired fellow agronomist José Guimarães Duque to oversee the nurseries. The following year, Trinidade turned several of Duque's nurseries into agricultural posts to support farming in their parts of the sertão. The posts were theoretically dedicated to the promotion of irrigated agriculture, though their principal activity was distributing seed samples. By 1937, four primary and eight subsidiary posts were operating across seven northeast states, providing sertanejos with instruction in agronomy, zootechnics, horticulture, and arboriculture. The largest four were located at Condado and São Gonçalo in Paraíba and at Icó and Lima Campos in Ceará; the smaller posts employed only one agronomist who gave advice, rented livestock and machinery, and sold insecticide at cost (repaid by farmers after the harvest).

Trinidade viewed irrigation as the second most important of IFOCS's tasks, after water storage in reservoirs (the most urgent necessity to sustain humans and livestock). He believed that irrigation was essential to stabilize the sertanejo population, providing them with greater food security and more varied economic opportunities. In the sertão's perennially warm climate, irrigation could allow for year-round cultivation, a tremendous boost to the economy. Trinidade recommended that irrigated farming operate in two different modes, for normal and drought years.[4] During droughts, irrigated fields would be used to grow corn, beans, and other crops basic to human sustenance. The residual material from those crops would sustain cattle as well. During normal years, when staple crops were abundant, resource-intensive irrigation could be more profitably used to grow arboreal cotton. Reliable soil moisture guaranteed a higher quality crop than could otherwise be grown in the sertão.

Trinidade felt that IFOCS should help develop a rational farming regime adapted to the sertão's environment to remedy centuries of inappropriate practices by sertanejo farmers. In his view, the region's relatively favorable climate during ordinary years disinclined farmers to invest in irrigation systems that they actually needed in order to survive droughts. Sertanejos were similarly complacent about making hay for cattle, since other forms of forage were normally plentiful. Farmers were also deterred from building their own irrigation systems by the dearth of perennial rivers. In the absence of a native irrigation tradition, the burden of developing irrigation networks fell heavily on the government. In contrast, other South American states that committed to irrigation projects in this period, such as Argentina, Chile, and Peru, were able to expand on privately developed systems.[5]

Trinidade was adamant that sertanejo farmers required education for irrigated cultivation to succeed. "The mission of the Northeast's agricultural posts is fundamentally one of education," he wrote in 1937. "Once the great dams are erected and the irrigation canals open, a huge challenge will still remain for the sertão—that of education."[6] In the agronomist's opinion, irrigated settlements in the American West had been less productive initially than they should have been because the U.S. Reclamation Service did not provide farmers with sufficient training. In contrast, several of IFOCS's agricultural posts included schools, cooperatively funded by the states, to help the children of farmworkers and landowners adapt to new ways of farming. The goal of these schools was to democratically "educate rural boys and girls, children of *fazendeiros* (landowning farmers) or of poor field workers, to live in a renovated sertão."[7] Their instruction would "cultivate a new attitude that knows how to make use of all the gigantic efforts which the Government of the Union is undertaking throughout the Northeast."[8] Trinidade reported favorably on visits by landowning families to agricultural posts, claiming that they returned home with renewed enthusiasm for getting the most out of existing reservoirs. Employees of the agricultural posts also became familiar with new ways of farming and living in the sertão. Instruction was always practical and "adopted to the mentality of the illiterate."[9]

In 1934 the Comissão Técnica de Reflorestamento e Postos Agrícolas do Nordeste was renamed the Comissão de Serviços Complementares da Inspetoria de Secas (CSC; Commission of Services Complementary to the In-

Agronomist leading a practical class on the use of a plow for adult workers at the São Gonçalo agricultural post, Paraíba. Source: *Boletim da Inspetoria Federal de Obras Contra as Secas* 11, no. 2 (1939): n.p.

spectorate for Droughts), to deflect the Ministry of Agriculture's interest in administering it. The CSC cooperated with the ministry's Department of Animal Production to evaluate livestock species best adapted to the sertão and promote animal husbandry suited to a semiarid climate. CSC agronomists also studied forage crops, since high cattle mortality during droughts typically triggered mass exodus from the sertão due to sertanejos' dependence on beef as a reserve protein source.

One of the CSC's educational practices was encouraging sertanejos to consume greater quantities of fruits and vegetables. These were not typically a significant component of local diets, largely because their cultivation had been unreliable. The CSC's agricultural posts experimented with produce appropriate to the sertão's climate and distributed seeds of successful specimens; they also made varieties of trees tested at the posts available to farmers for shade, soil retention, and beautification. Trinidade saw increasing the cultivation of fruit and shade trees as a way to encourage wealthier landowning families to begin residing in the rural sertão rather than in nearby towns. He hoped this would increase their personal and financial commitment to the area.[10]

Trinidade had more ambitious goals for the CSC as well. He described it as responsible for studying the impact of irrigation on the sertão's "agronomy, economy, and sociology," and aimed to affect sertanejos' lives well beyond IFOCS's planned irrigation basins. Ultimately, Trinidade aspired to tilt the balance of power in the drought zone away from estate owners by increasing the security of smallholders, who reaped fairly low yields per hectare without the benefit of irrigation and suffered from significant fluctuations in annual output depending on rainfall. "Only irrigation is capable of giving smallholders in the sertão the social and economic attitude [*sentido*] that they have in the South [of Brazil]," Trinidade wrote.[11] He also believed irrigation would only succeed if more sertanejos became smallholders: "Only the ownership of land by those who work it, or at least a long-term rental arrangement, will provide the farm laborer an incentive to take care of it. And without this, there is no possibility of irrigation. Land ownership is one of the central conditions of success for irrigated agriculture."[12] In Trinidade's view, large landowners, accustomed to independence and the exercise of personal authority, were not suited to cooperative use of irrigation networks. Estate owners were less adversely affected by discontinuities in crop production, since their overall yields were so great. They typically received one-half of the cotton harvest and one-third of the corn and beans from sharecroppers (*meieiros*) who worked their land. Droughts actually provided some benefit to wealthy *fazendeiros*, who took advantage of the sudden labor surplus to expand and repair their reservoirs, homes, and fences.

Trinidade was concerned that even when landowners did adopt irrigation they would not use it to cultivate crops beneficial to the sertanejo diet, such as fruits and vegetables; typically they focused on producing more profitable export crops. He was particularly opposed to employing irrigation for the cultivation of sugarcane, which he described as a *cultura individualísta*, ecologically and socially harmful. Nevertheless, Trinidade adopted a politic compromise between the interests of estate owners and their laborers, asserting that the goal of irrigation was to expand the sertão's economic base and not to supplant existing forms of production. Presumably CSC support for smallholding would eventually reduce the supply of dependent workers and thus increase labor costs—but this is not something Trinidade drew attention to. Like Miguel Lisboa before him, Trinidade calculated that concessions to estate owners were essential for the drought agency's political survival He supported IFOCS's construction of reservoirs on private estates, conceding that this was essential to secure the survival of the owners' cattle and workers (likely valued in that order as assets— though Trinidade did not articulate this).

Trinidade's lofty goals for the CSC were not matched by irrigation canal construction during the years that he worked for IFOCS. In 1937, three agricultural posts operated canals irrigating a total of 172 hectares (425 acres). Three years later, an additional post had a modest thirteen hectares (thirty-two acres) under irrigation. By 1941, 501 hectares (1,237 acres) were under irrigation at the CSC's posts.[13] The modest achievements of IFOCS's agricultural service during Trinidade's tenure indicate his minor influence within the drought agency. It is not surprising that he was unable to accomplish more during the 1930s given the preference of IFOCS's administrators for reservoir construction and the implications for rural power brokers of turning sharecropping sertanejos into smallholders.

José Augusto Trinidade died of cancer in 1940 at age forty-five, and in his honor President Vargas created an experimental institute at the São Gonçalo post. Trinidade had lobbied for this, to conduct "original research in this precise environment" on the sertão's climate and soils and on drainage methods necessary to avoid soil salinization. The new experimental station was inaugurated in 1940 as the José Augusto Trinidade Agronomic Institute.

Trinidade's successor at the CSC was José Guimarães Duque. The son of landowners from Minas Gerais, Duque dedicated much of his career to soil conservation and identifying marketable xerophilous (drought-tolerant) plants that could thrive in Brazil's semiarid zone. Trinidade had hired Duque in 1933, and Duque served as temporary head of the agricultural service from 1937 to 1939 while his boss's health was in decline. During those years, Duque moved the CSC's administrative offices to Fortaleza, Ceará, for easier coordination with IFOCS. Duque subsequently served as secretary of agriculture for the state of Paraíba, then returned to direct the CSC in 1941.

Under Duque's leadership, the agricultural service had eight main divisions: agronomy (seed distribution, development of improved and new crop species, and agricultural extension); horti-pomi-silviculture (cultivation of fruit trees and xerophilous plant studies); zootechnics (establishment of pure-breeding livestock strains, forage studies, silo construction, and milk production); soils (mapping and drainage studies); a laboratory (analysis of water and soil chemistry, and studies of drainage methods to promote desalinization); phytopathology, ecology and botany (crop pest and disease studies, use of insecticides, hybridization of plant species for improved resistance, climatic and meteorologic studies, and development of an herbarium); external cooperation (an extension service intended to support irrigation, administer water use, rent machinery, and offer public lectures); and reservoir administration (fish supply, land leasing along reservoir margins, tree planting, and management of mechanisms for water control). The CSC also included a medical and social service that operated a modest hospital and dental clinic, provided vaccinations, and distributed milk for children during droughts. Additionally, the CSC offered hygiene classes for sertanejo families and a group similar to a 4H Club.[14] It was a pioneer in agricultural extension within Brazil; Mexico's irrigation commission engaged in a similar range of "social development" activities in roughly the same period.[15]

Early in his tenure as head of the CSC, Duque described several sertanejo character traits that he thought made irrigation hard to implement; in particular, he perceived them to lack the cooperative spirit essential for successful irrigation. (Notably, a sociological study published in 1948 identified an "exaggerated" social cohesiveness as one result of sertanejos' shared drought experience—the opposite of Duque's perception.[16]) Duque blamed the European element in most sertanejos' lineage for their individualistic tendencies. Environmentally adapted indigenous knowledge was lost as Europeans moved into the region, he asserted, believing that the índios native to the sertão were better adapted to semiarid conditions and more cooperative with each other than later settlers (although he cites no evidence of early irrigation systems to support this). Such generalizations about the contributions of Africans, Europeans, and "Indians" to sertanejo culture were common among drought agency administrators in this period. Some wrote highly conjectural, personal musings influenced by popular construction of essential African, Indian, and other (Dutch or Portuguese) ethnic characteristics without reference to scholarly data. Several blamed native influence for the practice of burning fields to clear them, which exacerbated soil erosion during drought. Thomaz Pompeu de Sousa Brasil Sobrinho, in an essay for IFOCS's quarterly journal, asserted that the sertão appealed to the descendants of Indians because it offered an independent life free from "administrative care." These views contrast with Duque's imagined communities of native Indian irrigators.[17]

Duque argued that sertanejos needed to be taught how to avert calamity through a process of guided acculturation, and he advocated formal and informal education to remake ordinary sertanejos into active "citizens of the sertão." At the São Gonçalo agricultural station, farmworkers' children spent three hours daily in a classroom followed by two hours of "practical education" in the fields. Then they worked on the post's farm to contribute to their families' income.[18] Duque used radio programs to publicize new agricultural techniques and methods of soil conservation to far-flung sertanejos; he promoted farming practices that made the most efficient use of available land, such as planting pasture below carnauba palms (a species cultivated in the sertão for its commercially valuable wax). Such bilevel cultivation had the added advantage of providing protection against wind erosion for both the pasture and the palm roots.[19]

Like Trinidade, Duque hoped that IFOCS's agricultural programs would increase the productivity and security of small farmers in the sertão. But during the 1940s he downplayed many of the political dynamics that impeded the expansion of smallholding—particularly elite opposition to state expropriation of land surrounding reservoirs, even if that land had long been unproductive. Duque often described sertanejos as if they were a uniform population, without acknowledging the significant economic and political differences separating estate owners from the landless. In his profoundly depoliticized framing, sertanejo farmers lacked the will and discipline to achieve economic independence. Their periodic misery was a result of weak moral character, he asserted, noting, "The hour has arrived for the people [o povo] to participate actively in the destiny of their environment, to help resolve, definitively, the questions that will determine the survival of all, and not continue to be mere spectators to the government's initiatives, accomplished through the work of the técnicos. It is impossible to overcome the climatic irregularity and the obstacles to agricultural production permanently with a group of people indifferent to the fortune of their environment, inactive and self-centered."[20] Duque's paternalistic assertion that sertanejos simply required tutoring in the moral and civic habits necessary for increased agricultural production reflect the conservative reformism typical of Vargas's administrations.

In the late 1940s Duque published a book that comprised his recommendations for farming in a semiarid environment. Much of his advice emerged from IFOCS agricultural service's research into dry farming, "the practice of agriculture without irrigation in regions of limited natural precipitation."[21] Dry farming entails selecting crops particularly suited to semiarid environments, alternating years of crop and fallow, and employing tillage techniques that conserve moisture, such as packing the subsoil to reduce evaporation. The technique had been developed by the U.S. Bureau of

Reclamation and promoted during the early twentieth century to attract settlers to the semiarid northern U.S. plains. Snow melt aided dry farming in semiarid U.S. regions, but this source of groundwater was not available in the equatorial sertão. Many dry-farming methods were tested through trial and error by U.S. settlers, whose efforts to farm western states often failed.

Duque's *Solo e Água no Polígono das Sêcas* (Soil and water in the drought polygon) recommends cultivating commercially valuable xerophilous plants (species that flourish in hot, dry climates) throughout the sertão, thus turning the region's climate from a liability into an asset. The plant species that he recommends include oiticica for its nut oil; manicoba for rubber; mangabeira for latex; licurizeiro palm for its wax and oil; murici, batiputa, and umbuzeiro for their fruit; faveleiro for its tasty seeds; carua for fiber; and manipeba for its starchy roots, storable for several years as food security against drought. Duque also lists more than one hundred varieties of palm that provide good forage for cattle. He argues that improving sertão agriculture is an important contribution to national industrialization, since the commercial crops produced by sertanejo farmers will become the bedrock of regional industry.

Duque published a second edition of his influential book in 1951, expanding beyond technical discussion of climate, soil, and conservation. The revised version emphasized the need to educate sertanejos about irrigated farming and accustom them to new farming practices. He accused sertanejo farmers of engaging in "a collective, organized movement to destroy the natural wealth that will affect . . . the livability and productivity" of their region by burning fields to clear them.[22] He argued that it was up to DNOCS's agricultural service to "prepare the common man to better take advantage of [engineering works to] complete the work of the *técnico* who abruptly introduced irrigation, like a wedge, into the Northeast's social organization."[23] Promoting habits of cooperation among farmers was as central to irrigation's success as any technical effort, Duque asserted: "Knowledge of engineering, botany, agronomy and medicine are not sufficient [for the success of drought works]. An elevated level of sacrifice, a deep human understanding of the population's needs, an almost messianic Christian spirit, an absence of egoism are all needed in order to resolve the drought problem. . . . The social aspect of droughts is equal to its technical aspects."[24] To fulfill its educational goals, Duque's Serviço Agro-Industrial operated seven schools at its agricultural posts by 1953, serving more than four hundred children. Classes were held outside, in gardens and fields, as well as indoors. Students learned reading, writing, and math along with economic subsistence skills, modern hygiene, and "goodwill to serve the community."[25]

Approaching agronomy as a civilizing mission, Duque portrayed sertanejos as fit for modernized production if they were provided with the appropriate scientific tools and education. For irrigation to succeed, farmers needed to develop a disciplined work ethic and learn to cooperate with their neighbors. Duque described irrigation as requiring "Christian virtue" while engaged in "the joint exploitation of a public reservoir by a group of families financed by a social drought fund; the rental of lots with control over their fertility exercised by an agronomist representing the government; the sale of harvested crops coordinated under the direction of a private association; and the collective purchase of materials and necessary items by the same administrative organization: these are the only ways to avoid the exploitation of man by man and to maintain production throughout generations."[26] Like Trinidade, Duque focused on smallholders as the most likely to adapt to irrigation's demands, since they suffered so much during droughts. Although migration to less affected parts of the sertão had been a way for families to escape droughts, this became less practical as the sertanejo population grew. Duque estimated that a family could reliably subsist by farming ten hectares of irrigated sertão land and one hectare within a reservoir basin when its water was low (called *vazante*, or "ebb tide," cultivation). But many small properties were not well laid out for irrigation, since they extended in narrow strips perpendicular to rivers—the result of larger properties having been divided over generations to comply with partible inheritance laws.

During the 1942 drought Duque persuaded landowners in the irrigated basin surrounding the São Gonçalo agricultural post to allow migrants to farm irrigated plots on their property, since the owners themselves had shown little interest in making use of the post's machines, canals, and extension services. More than eight thousand people from 1,460 families were accommodated in this way, on 1,122 hectares of land. (Due to the Allies' interest in rubber for their war effort, many drought migrants were relocated that year to the Amazon and employed as rubber tappers; Vargas's propaganda machine euphemistically termed them "soldiers of rubber."[27]) A decade later, during the 1953 drought, five thousand irrigated hectares were available for migrant families to farm around São Gonçalo.[28]

In response to that devastating drought, Duque published a final edition of *Solo e Água no Polígono das Sêcas*. Here he made the most forceful case for the need to transform the sertão—both socially and technologically.

Regionally appropriate education was essential, he insisted, to take advantage of the "robust but latent intelligence in the soul of the masses."[29] Adopting an ambitious view of his agricultural service's mandate, he argued that expropriation of land from estate owners, and rental by the state of irrigated plots on that land to farmers, was essential for both soil conservation and sufficient food production. Estate owners typically devoted their best land to commercial crops, which left only the less fertile soil for food cultivation and contributed to its depletion. By settling poor families around publicly funded reservoirs, Duque's Serviço Agro-Industrial would guarantee food sufficient for the growing sertanejo population (12.5 million in 1950 and increasing by 2.4 percent annually). This, he hoped, would minimize outmigration during droughts, which damaged "the state, society, and the family" by draining young, productive men into urban capitals that did not readily absorb their labor.[30] In this way, Duque described land expropriation—a politically volatile proposal—as having technical and economic motivations as much as social aims. Large cooperatives of smallholding irrigators would combine the economic advantages of *fazenda* estates with the social advantages of *minifundia* (smallholding), he argued. As the sertão's population grew, new colonies could be established along the more humid western periphery of the drought zone, particularly in Maranhão, with roads linking their products to northeastern markets. Agronomists would "establish medical and religious assistance, civic education and hygiene, and technical agricultural instruction" at these settlements, helping farmers learn to relate to each other as plants do in a harmonious ecosystem.[31]

The third edition of Duque's book emphasized the need for improved government administration to oversee northeastern development. Duque underscored that successful irrigation systems require substantial investment in soil and geographic surveys prior to canal construction, which DNOCS needed the means to afford and conduct effectively. But Duque did not place sole blame for the failure to irrigate the sertão on the government; sertanejos of all classes also had to recognize their stake in the region's future and take full advantage of the drought agency's investments. According to Duque, half of the private land on which DNOCS had constructed irrigation canals remained uncultivated in 1953.[32]

## Agronomists' Enthusiasm for Smallholder Irrigation

Many Brazilian agronomists working for the drought agency in the 1940s were optimistic about the dramatic social transformation that could come from expropriating land around government-owned reservoirs to form irrigated colonies. One reported to IFOCS in 1941 on the exploitative labor conditions endured by tenant farmers living near the Forquilha Reservoir and its agricultural post (eighteen kilometers from Sobral in the state of Ceará). On unirrigated farms, tenant farmers (*colonos*) were paid a daily wage of Rs2$000–2$500 (2–2.5 mil réis),[33] equivalent to the cost of one liter of rice, corn, beans or milk, or less than one kilogram of meat. They worked three days each week for the landowner, and the rest of the week in their own small gardens (*roças*). *Colonos* on irrigated plots retained half of their harvest, which included sugarcane, bananas, corn, and beans; the landowner provided seeds, while families supplied labor for clearing, planting, irrigating and harvest, using axes and hoes. The owner generally lived most of the year in a nearby city, engaged in other business, but maintained a comfortable home on his rural property, constructed of brick with a tile roof. His workers lived in houses made by hand, with walls of wooden sticks and pounded clay (wattle and daub), and a roof of palm thatch; these were, according to agronomist Inacio Barreira, "small, low, and dark without comfort or hygiene, in keeping with the indolence, rusticity and degree of backwardness, ignorance and lack of education of their inhabitants."[34]

Typically the only furnishings were kerosene cans used as chairs. Although two schools were located in the municipality, funded by the local and state governments, they lacked desks and teaching materials, woefully limiting their teachers' effectiveness. Barreira asserted that land expropriation by the government to create an irrigated colony of smallholders would significantly improve rural sertanejos' economic security, and he embraced the assumption, widespread among progressive reformers, that improved education and acculturation to more disciplined work regimes were central to regional economic uplift.

Agronomist Trajano Pires da Nobrega, analyzing the economy and population along an extension of the São Francisco River, twenty-five kilometers from Itaparica in Pernambuco (the terminal point of the Paulo Afonso train line) in 1941, proclaimed that an irrigated smallholder colony there would "initiate the formation of new Brazilians, namely these transformed families."[35] In his report for IFOCS, Nobrega described the modest local economy in which most crops and livestock were cultivated for domestic consumption. Many manufactured items and even basic foods had to be imported, often from Recife. These included *rapadura* (cakes of raw sugar), salt, butter, clothing, hammocks, kerosene, soap, and shoes. The unnavigable stretch of river at the Paulo Afonso waterfalls cut the area off from easy transport of

Home built by a worker at the São Gonçalo agricultural post, Paraíba: a temporary construction "in accordance with the nomadic spirit of the drought migrant" (according to agronomist José Guimarães Duque). Source: *Boletim da Inspetoria Federal de Obras Contra as Secas* 11, no. 2 (1939): n.p.

goods to or from surrounding communities. A postal bus traversed part of the region weekly; the remainder was served by an unreliable mule post that covered a two-hundred-kilometer route. Many farmers did not have documented title to the land that they claimed ownership of based on long-standing occupation and cultivation. Uncertain rain discouraged agricultural investment except along the riverbank, where vazante cultivation was reliable in the evaporated riverbed even during dry seasons. Along the river most did own their land, "pulverized" through generations of partible inheritance into farms sometimes three by forty meters in dimension.

Nobrega viewed the inhabitants optimistically, as extremely independent and hardworking farmers and ranchers.[36] He describes them as "*caboclo* [mixed-race native-European] types" who inherited the "independence and lack of discipline of their indigenous forebears" but for whom the strong desire to own and improve their land had resulted in "genuine evolution that should form an excellent basis for selecting future irrigators."[37] He deemed the residents to be physically robust, but they were below average in height

Sertão home of a relatively affluent *coronel*, his wife, and their nine children, 1912.
Source: Acervo da Casa de Oswaldo Cruz, Departamento de Arquivo e
Documentação.

due to nutritional deficiencies. Like Duque, Nobrega thought the greatest
impediment to the success of irrigated farming was the prevailing individualism and suspiciousness (*desconfiança*) of local farmers and that agronomists needed to instill a cooperative spirit within the community. He hoped
that the wealth generated by initial irrigators would break other farmers'
distrust of novelty and impel them to join the project. Thus it was important to focus on easily harvested crops with high economic value, such as
rice, beans, corn, manioc, sweet potatoes, cotton (in modest quantity), and
alfalfa for cattle forage. A secondary priority should be improvement of the
local diet, expanding fruit cultivation and raising dairy cows to produce butter and cheese. Nobrega also recommended "introducing one or two foreign families, preferably from regions best known for their agricultural
capacity, like Germans or Poles," to model profitable irrigated farming. This
was a common suggestion among mid-twentieth century sertão developers,
reflecting prevailing assumptions about the cultural and racial superiority
of Europeans.[38]

Nobrega proposed that an initial trial be conducted at an agency agricultural post. Plots could be rented to the post's workers, guided by agronomists

in cultivation techniques, use and repair of machinery, methods of crop transport, and home economics—the latter so they would not squander their profits, thereby jeopardizing the value of the experiment as a model for other farmers. Undisciplined settlers would be eliminated through this process, but after three years successful ones could move to their own plots in an area of the river valley acquired for this purpose by the federal government, via land expropriation. Nobrega estimated that if every family in the new settlement received five hectares of irrigated land and two to three dry hectares for livestock, a three-thousand-hectare stretch of river in the Chapada Valley could accommodate five or six hundred families. Along with the regional economic benefit that such a scheme could provide, it would retain sertanejos during droughts (temporarily accommodating up to five thousand people, and thousands more cattle), reducing their undesirable migration to São Paulo. Ultimately, Nobrega envisioned a city emerging around the irrigated colony, providing modern amenities like electricity, water, and sewage services; a well-appointed school; a cinema, market, and health clinic; and clean parks and squares—in short, "a center for credit, production, and consumption."[39] Private investors, perceiving the unfolding commercial opportunities, would fund further irrigation canals, accelerating the region's economic development. "This will be the process by which the sertão of the São Francisco [River] valley is transformed into the promised land," Nobrega concluded grandly, building on the asset already provided by IFOCS's road network through the region. "This is not a utopian plan," he insisted. Yet, judging by the drought agency's quarterly bulletins, few such settlements were actually established during the 1940s outside of the agricultural commission's own posts, due to the professional priorities and political calculations of engineers who managed the drought agency.

During the 1950s several agronomists published training manuals for new sertanejo irrigators. An undated, handwritten pamphlet illustrated with simple line drawings, *Irrigante amigo! Seja bem vindo ao Projeto São Gonçalo* (Irrigator friend! Welcome to the Project São Gonçalo) presented settlers with a list of their contractual rights and responsibilities.[40] From DNOCS they would receive technical assistance, year-round water, medical and dental care, schooling for younger children, financial credit, and membership in the post's agricultural cooperative; in return, they were to use their lot for farming and ranching and maintain good relations with surrounding families. Irrigators should accept agronomists' guidance about what crops to plant and how to cultivate and market them through the cooperative. They were to work hard, maintain their canals, drains, and homes, and participate in meetings, classes, and social activities sponsored by the management. "If you follow these instructions, you will be a good *colono*," the pamphlet concluded. A more formal publication along the same lines specified that the irrigator's primary goals were to avoid soil erosion and salinization and to maximize his family's income by following agronomists' advice regarding crop and seed selection, timing of planting and harvest, fertilizer use, pest removal, irrigation methods, and crop storage.[41]

Carlos Bastos Tigre's moralizing *Catecismo do Agricultor Irrigante* (Irrigator's catechism) was published by DNOCS in 1954 as a wide-ranging advice manual for participants in the agency's irrigated settlements. Tigre encouraged families to construct their homes of cement, tile and wood, use "modern and elegant" kerosene or electric lamps, install piped water and sewage pipes leading to a cistern or septic tank, construct separate housing for livestock, hang attractive curtains, and place perfumed plants and fruit trees around their property. "If a farmer pastures cattle on lands served by irrigation canals, this shows that he is ignorant, has no sense of responsibility, no spirit of cooperation, is an egotist and unpatriotic," Tigre admonished.[42]

He encouraged irrigators to seek a range of advice and services from DNOCS's agricultural posts. By cooperating with each other they would foster "community progress and national prosperity," demonstrating their gratitude to the government for providing them with valuable infrastructure and assistance.[43] Tigre's catechism extended to nutrition, advising settlers to eat fruits and vegetables as well as fresh eggs, meats, and fish of known provenance, and warned that "a bad diet saps your strength, disposition, and initiative."[44] He also offered guidance about personal hygiene, pregnancy, and sanitary infant care, opining, "The death of so many children in the sertão is due to poor food hygiene and their parents' lack of the most basic understanding of nutrition."[45] This is a strikingly depoliticized view, relative to the contemporaneous analyses of the economic structures underlying malnutrition by Pernambucan doctor Josué de Castro.[46] Tigre encouraged moral discipline of adolescents, especially boys, so that "abuse of their bodies" would not lead to imbecility, insanity, or venereal disease; he claimed that "they will thank you and God when they reach maturity for having given them such protection."[47] The sertão's 1950s modernizers clearly understood their educational purview to extend well beyond agricultural practices.

By 1958, under President Kubitschek's aggressive program of infrastructural expansion and industrialization, DNOCS oversaw 177 public reservoirs holding more than 6.4 billion cubic meters of water. More than 65,000 people were living in irrigated reservoir basins administered by the drought agency, and a thousand students participated in schools or farming clubs at agricultural posts. Yet less than 8 percent of the sertão could be irrigated, requiring alternate productive use of the remainder, particularly through dry farming and ranching.[48] Agronomists continued to proselytize rational use of Brazil's semiarid land, ecologically adapted for maximum crop yield, economic benefit, and enjoyment. Progress along these lines was limited not by natural or financial resources, according to Duque, but by "incapacity to cooperate and elite egoism" combined with a lack of political will to prioritize the economic needs of the rural poor.[49] Many public works in the sertão remained underutilized for these reasons. The need to increase food supply and pasture grew more acute as the population continued to grow— despite periodic outmigration. Duque remained confident that with strategic use of seed selection, irrigation, dry farming, fertilizer, pesticides, intercropping, and rotation of human food and livestock forage, sertão farmers could adequately feed the region's growing population and increase their per capita income. Even so, he feared cultural impediments to such progress: the reluctance of young men to subordinate their capricious whims to the discipline of intensive cultivation, and the tendency to waste profits on ostentatious trinkets or alcohol rather than investing in property, education, and agricultural tools.[50] In a 1959 lecture to Rio's engineers club, Duque remarked that much of the agronomists' work required gaining farmers' trust and persuading them to break with their communities in the interest of regional progress.

## Agronomists on the Front Lines of Drought Aid

Archival records of day-to-day work on the drought agency's construction sites in Ceará (the only state for which such records remain available) during the 1940s and 1950s reveal agronomists to have promoted both land redistribution and the acculturation of poor sertanejos to more productive cultivation practices. Their vision for sertão transformation entailed restructuring the material basis of political power (insofar as regional dynamics allowed) while simultaneously reforming the most marginal farmers into disciplined, productive workers. To accomplish this, agency agronomists tried to combat both elite corruption and self-interest and the apparent recalcitrance of the poor whom they intended to help. They engaged in the "middle politics" described by Michael Ervin with regard to Mexican agronomists in the 1930s. Under President Lázaro Cárdenas's revolutionary administration, Mexican agronomists promoted land reform, which pitted them against elites; yet they had difficulty persuading *campesinos* (peasant farmers) to provide census data (for fear of tax implications) or to embrace nationalist motivations for intensifying agricultural production.[51] Sertão development also involved multilevel negotiations, including those between representatives of government agencies and the "targets" of their assistance. Local farmers often contested reformers' recommendations for more efficient and productive land use.

Conflict between owners of land surrounding reservoir basins and self-identified small farmers who worked that land are evident in a petition from a syndicate of men living around Choró Dam in Ceará that was submitted to the federal labor ministry in 1939.[52] (Choró Dam had been constructed as a "work front" in 1932–34 under Brasil Sobrinho's direction.) Detailing the various injustices that they suffered at the hands of exploitative landowners (who demanded half of their crops, or required several days of work each week in return for the rental of modest plots, and who charged unjustifiably high prices for basic necessities in their local store), members of the syndicate asked why such powerful men were *themselves* allowed to rent land around the public reservoir. The farmers petitioned the labor ministry to grant them a range of worker protections already legislated under President Vargas for urban industrial workers; these included a minimum wage, paid holidays, and disability compensation. (Rural workers would not receive similar protections until the early 1960s). They also desired dispensation from licensing requirements for fishing in the reservoir, which was necessary to sustain their families. Finally they requested seed provision and equipment loans to help their cooperative become more productive.

Responding to the Ministry of Labor, Industry, and Commerce's (Ministério dos Negocios do Trabalho, Indústria e Commércio) request for further information about this conflict, drought agency staff working in the area confirmed the accuracy of *sindicato* members' grievances. The technical director for Choró explained that property owners who "donated" land to the federal agency during the expropriation process that preceded reservoir construction—for which they were typically compensated at 25 percent of market value for their home, fields, valuable trees, and wooden fences—had the right to rent a small plot of irrigated land near the reservoir. He himself deemed this policy "antisocial and antieconomic," counter to the agency's goal of promoting more equitable land distribution in the sertão. Yet a later communication to the drought inspector from other agency staff at

Choró cautioned that the landowners around the reservoir were poor farmers themselves, hardly *commerciantes* or *capitalistas*.[53]

Oversight of the Choró Reservoir and its irrigation canals, which watered approximately 950 hectares (roughly 2,300 acres), passed from the drought agency's engineering division to its agricultural service in the 1940s. Agronomists' initial survey of the project revealed various irregularities in landholding arrangements around the reservoir. Many irrigators who rented plots from the agency were in flagrant disregard of legislation limiting each family to ten hectares of *terras secas* (dry land) for their home, livestock, and less water-demanding crops and four hectares within the irrigated basin for food cultivation. Preference was supposed to be given to the poorest applicants. Yet some settlers had accumulated up to twelve times the legal limit for property rental, distributed among extended family members. These families often sublet parcels to sharecroppers or wage laborers, including drought migrants who were supposed to be the primary beneficiaries of reservoir construction. This corrupt system replicated the social organization that sertão reformers aimed to displace. The drought inspector explained to the minister of public works that his agency was rescinding several contracts of families who had accumulated unjustifiably large parcels, to "rescue" 282 families (1,270 people) from the "inhumane" means of sustenance in which they were presently engaged. This would deliver on DNOCS's mandate to serve "innumerable poor workers who genuinely need it" and to increase food production, creating a "more vibrant outlook" in the sertão (*perspectívas mais animadores*).[54]

Not surprisingly, many renters of irrigated plots protested vigorously against the rescinding of their contracts. José Delfino de Alencar, who effectively controlled 122 hectares of dry land and fifty-eight vazante plots within the reservoir basin (accumulated in the names of his children) successfully sued the drought agency in court. Duque and his engineering colleagues bitterly disputed this ruling (which was appealed multiple times), calling the outcome "a matter of life or death" for several poor families who would be able to farm the contested property if it were wrested from the Alencar clan. Duque's supporters in the drought agency depicted such conflicts as a struggle over the civic health of the sertão, pitting progress against "begging, affliction, crime, and communism."[55]

Many drought agency agronomists hoped to mitigate inequalities in the sertão through their infrastructural projects. Despite these altruistic motivations, there is substantial evidence of conflict between agency personnel and the irrigators they intended to help. Generally, irrigation was desired by sertão farmers during droughts but was rarely used or maintained in years when it was not essential for survival. In one instance, the director of the agricultural post around Forquilha Reservoir complained to Duque about farmers' use of irrigated plots for pasture. This damaged both canals and drainage systems. The "recalcitrant" livestock owners apologized when agency staff brought this problem to their attention, but did not change their behavior—in short, the *cooperantes* (as irrigators were called) "weren't co-operating!" Citing such cases, Duque petitioned the Ministry of Public Works to promote legislation that would expropriate all land crossed by federal irrigation canals, along with one hundred meters on each side to act as a buffer against roving cattle. This would give the agricultural service a stronger legal basis for punishing those who abused their infrastructure, much of which had been constructed on private land.[56]

Conflict between agronomists and irrigators escalated around Forquilha over the 1940s. At one point, water serving the property of Silvestre Gomes Coêlho, who allowed his cattle to roam across neighboring irrigators' land, was turned off. He was fined for damages to the canals, which were deemed national property. In response, Gomes Coêlho brandished his rifle at agency staff who arrived to halt water flow to his property and threatened to blow up the water meter if his irrigation access was not restored. In Duque's assessment, such "acts of low education" (meaning low social class) were typical of this fellow, who could rarely be persuaded to follow "rules of good comportment" (*regras do bom caminho*). The agronomist stationed at Forquilha deemed such behavior "unconscionable," especially when engaged in by the very sertanejos for whose benefit the nation had constructed and maintained irrigation networks at great effort and expense.[57] From the mid-1940s onward canal restoration became a reliable source of employment for drought migrants due to the significant damage caused to them by free-ranging livestock.

Eight years after the fraught interaction with Gomes Coêlho, Duque was mired in debate about how much water could be drained from the Forquilha Reservoir by irrigators. The basin also yielded a three-thousand-kilogram annual fish harvest, providing sustenance for nearly forty thousand people in the area.[58] Due to a drought in 1953 and resulting heavy water use by irrigators, reservoir volume had fallen to 5 percent of capacity. By 1957 only one-fifth (100 hectares) of the land served by irrigation canals could be farmed due to diminished water volume and leakage from canals that were in poor repair.[59] As examples of this kind demonstrate, drought agency staff were frequently embroiled in disputes over which populations would benefit from their management of land and water and which local residents would bear the cost of this restructuring. As early as the 1920s conflicts had arisen about how much water the drought inspectorate could store in its reservoirs and how much should be allocated for different agricultural uses. At that time Minister of Agriculture Ildephonso Simões Lopes directed a commission to rewrite Brazil's water code, drawing on U.S. legal policy in the American West, where property owners were permitted to diminish the quantity of public water flowing through their land if they put that resource to productive economic use.[60]

## Popular Views of Mid-Twentieth-Century Drought Works

The folk poems know as *cordéis*, sold in northeast Brazilian markets since the 1860s cotton boom, provide a rich source of popular commentary on the drought by people who are poorly represented in official records of regional development. Cordéis cannot be interpreted as revealing any uniform sertanejo response to political events (particularly from the 1950s on, when poems were sometimes commissioned by candidates to influence public opinion), but they reflect a spectrum of Nordestino opinion about a range of subjects over the twentieth century. Prolific Bahian *cordelista* Rodolpho Coelho Cavalcante asserted that successful poets must be attuned to the marketplace, so their products often echo (and simultaneously influence) public opinion.[61] Some critics of sertão development policy recommend cordéis as an aid to comprehending not only the lived experience of drought and migration but also the shortcomings of government response.

A few cordéis target IFOCS engineers as participants in the political corruption that confounded twentieth-century drought aid. In an undated poem probably from the 1930s, a master mason and his friend visit the chief of drought works in their area to ask for employment. They bring a recommendation letter signed by an influential patron, but the boss ignores their credentials and says he can tell they are "vagabonds, imbeciles without a profession . . . dreaming of the sertão." When their pleas for work relief are disregarded, the mason grumbles that, given how things operate in the sertão, it would be better to be a thief than an honest worker. The chief grows angry at this and tells the two what he knows of thieves from when he sold his land for almost nothing to an unethical middleman following drought in 1922. He had sought help from the same patron whom the mason offered as a reference, but that influential man provided no help, being caught up in parasitic land speculations himself.

The narrator of this *cordel* refers to the embittered engineer as one of Brazil's "financial dreamers" who believed that drought works would bring wealth, although they appear to be nothing but a colossal expense:

| | |
|---|---|
| *Relativo aos grandes gastos* | As far as the enormous sums |
| *(No tocante as açudagens)* | (Referring to the dam reservoirs) |
| *Os resultados são parcos* | The results are scant |
| *São simples sonhos miragens!* | They are simply dreams, mirages! |
| *Somente os debitos da patria* | Only the country's debts |
| *São negras grandes imagens!*[62] | Are big, black images! |

The mason concludes that his only hope is to die during the drought and finally be "free from engineers / who are plague, hunger and war" (Livre já dos engenheiros / Que são peste, fome e guerra). This cordel author, J. Evilasio Tavares, courageously named the local dignitaries whom he viewed as friends or foes of ordinary sertanejos in the face of drought.

The ineffectiveness and corruption of state authorities, and the scarcity of funds that actually reached the poor once they were remitted by the federal government, are recurrent themes in drought cordéis. Raimundo Santa Helena explains the problem bluntly in his 1932 cordel "Flagelados das Secas," having suggested the sort of productive projects that the government might engage in to genuinely assist the starving:

| | |
|---|---|
| *O Poder, que é mau político* | Those in power, who are lousy politicians |
| *Não tem medo do abismo* | Have no fear of hell |
| *Vê no pobre um paralítico* | They view the poor as paralyzed |
| *Vê no rico o altruísmo!* | They view the rich as altruistic! |
| *Dividir terra é fatídico* | To divide land is a tragedy |
| *Ao feudo coronelismo,* | To feudal *coronelismo*, |
| *Que na seca é urubu:* | Which during drought is an undertaker: |
| *Com açude, tem fartura—* | With the drought works come abundance— |
| *Compra um jumento, usura!* | Buy a redneck, avarice! |
| *Com um prato de angu . . .* | With only a plate of *angu* . . .[63] |

Santa Helena lists all the relief works that the government should undertake: building reservoirs, providing farming tools and seeds, improving fish stock, planting drought-resistant trees, and providing silos for crop storage. If not, he warns, "The Northeast will become a desert, burying its skeletons." Many of the projects that the poet recommends were promoted by

Minister of Public Works José Américo Almeida at the time, so Santa Helena may have intended to express support for Almeida's agenda.

Not all cordéis characterize drought aid as corrupt or mismanaged. Some praise the generosity of particular potentates for their administration of charitable assistance while criticizing other leaders in contrast to those admirable few. A 1922 poem, "A Ceca [sic] do Ceará," extols the governor and elites of the state of Sergipe for accepting a shipment of *retirantes* (drought migrants) from Ceará.[64] Another cordel written thirty years later salutes a Paraíban bishop's radio call for solidarity with the drought victims, contrasting the success of his appeal with local police ineptitude.[65] Bahian *cordelista* Cavalcante became an unabashed propagandist for President Vargas's drought alleviation efforts in several undated cordéis, probably written during the early 1950s. Recognizing sertanejos' desperation, Cavalcante noted, the merciful president had determined that aggressive steps must be taken in the form of reservoir, road, and railway construction to provide temporary employment. In Cavalcante's idealized depiction, sertanejos responded industriously, trading their labor for food, donated clothing, and medicines. The poet praised Vargas's patriotic devotion to his compatriots ("devoção e brasilidade") and heaped blessings upon him. Cavalcante finished by identifying himself as an untutored *propagandista* and vindicator of the poor, offering inspired verse for their entertainment.[66]

From a contemporary perspective, the most jarring cordéis blame the suffering caused by drought on sertanejo mores. One pair of nearly identical cordéis, both titled "Retirante" (the latter presumably plagiarized from the former), emphasize the fecundity of sertanejo families and the resulting difficulty of obtaining sufficient food during drought. As the story is recounted, a family of more than twenty children arrives at a sugar mill, pleading for work and food. The father claims to have lost more than half of his offspring en route, and his wife is evidently pregnant. Their tale becomes stranger: the wife gives birth during the family's exodus, aborts another pregnancy, and pleads for assistance from her patron saint, Padre Cícero (an uncanonized but beloved *cearense* priest).[67] This poem, republished in Padre Cícero's native town of Juazeiro, Ceará, criticizes sertanejos as irrational and superstitious—traits that left them at the mercy of others' benevolence.[68] The *cordelista*'s views reflect the growing concern with overpopulation that was internationally pervasive by the 1950s, though disputed by prominent Brazilian intellectuals like physician Josué de Castro, who instead viewed structural inequities as the root cause of food scarcity.

Another form of sertanejo blaming, one with biblical overtones, identified the population's lack of virtue as the cause of drought and consequent suffering. One 1953 cordel depicts drought as a curse leveled against female immodesty and hedonistic culture. Severino Borges Silva begins his poem in a sympathetic vein, describing the desperation of drought migrants in two cities and the generosity of those who follow Christ's injunction to feed the sick and clothe the naked. In that year food crop harvests were insufficient in several southern states as well as the sertão, leading to severe scarcity and inflated prices nationwide. Borges Silva's poem describes people forced to eat clay and sand, with thousands on the brink of starvation. The narrator warns that this misery is a punishment from God: "Tudo é castigo que Deus / manda para o pessoal / que vive de farra e dito / e festas de carnaval / e sem se lembrar que tem / um poder celestial" (God sends only punishment / to the people / who live for pleasure / and the festivals of Carnaval / without remembering that there is / a celestial power."[69] The poem chastises young girls: "([they] paint their lips and nails, running around at all hours; this is why God punishes [us] / and nothing is improving." Despite the admonitions of fathers, young women continued their wayward behavior, and this had caused tidal waves, earthquakes, and drought throughout Brazil. The scourges offered a moral lesson:

| | |
|---|---|
| Pois já está chegando o tempo | So the time is arriving |
| Do povo sofrer na terra | When the people suffer on earth |
| Castigos e mais castigos | Punishments upon punishments |
| Prá ensinar a quem erra | To instruct those who err [sin] |
| Com doenças perigosas | With dangerous diseases |
| Sêca fome pesta e guerra | Drought, hunger, plague and war |
| | |
| Pelos Estados sulistas | Throughout the southeastern states |
| Se acaba todos os dias | People are dying each day |
| Gente de fome e de febre | From hunger and fever |
| E outras epidemias | And other epidemics |
| São os principios das dores | These are the onset of the suffering |
| Como diz nas profecias | Described in the prophecies |
| | |
| Porisso eu aviso a todos | Therefore I advise everyone |
| Deixem uso farra e dito | To desist from pleasure-seeking |

| | |
|---|---|
| *E vamas pedir auxilio* | And let us beg for aid |
| *Ao nosso pae infinito* | From our infinite father, |
| *Que ele não nos socorrendo* | For without his help |
| *Não fica vivo um mosquito. FIM* | Not [even] a mosquito will remain |
| | alive. END. |

A similar explanation was offered in several cordéis describing torrential floods, including the deluge caused by the Orós Dam break in 1960. In one, not only the water's destruction but also inflation and poverty are blamed on the ambitiousness and moral laxity of modern society.[70] Such interpretations indicate the authors' unease in the face of Brazil's mid-twentieth-century economic and cultural modernization.

· · · · · ·

Folk music provides another window into popular responses to drought works, particularly during Vargas's administrations. During the authoritarian Estado Novo of 1937–45, Vargas's Ministry of Education and Health vied with his Department of Press and Propaganda for influence over national culture.[71] Radio programs were one way in which the federal government tried to shape national conceptions of *brasilidade* (Brazilianness) in the mid-twentieth century. In the late 1940s a sertanejo musician named Luiz Gonzaga became a national sensation through programs sponsored by the state-owned Radio Nacional. Gonzaga was a man of medium-dark complexion, born in the interior of Pernambuco where he played accordion with his father for local festivals. He was portrayed as an "ambassador from the sertão" who had come to introduce other Brazilians to the customs of his homeland.

Gonzaga had joined the army in 1930, which had given him an opportunity to travel around Brazil. Later in that decade he worked as an entertainer in Rio de Janeiro bars. By the mid-1940s Gonzaga and his *cearense* collaborator, Humberto Teixeira, were performing traditional and new sertanejo songs on Radio Nacional. From 1950 to 1951 they presented a weekly radio program, *No Mundo do Baião* (In the land of baião) that featured *Nordestino* customs and folk humor interspersed with musical numbers.[72] Gonzaga sang in an exaggerated regional accent, attired in the costume of the sertão's notorious *cangaceiro* bandits (whom, ironically, his battalion had helped Vargas to eliminate). He promoted a northeastern musical genre called *baião*, and his songs quickly came to represent authentic sertanejo culture for many Brazilians. The lyrics that Gonzaga popularized describe the beauty of the northeastern interior and its heroic people: fathers protecting their family's honor, loyal women waiting for their lovers' return, celebrants at weddings and harvest festivals, and famous outlaws from the region's turbulent history. The show was sponsored by a packaged food company catering to modern Brazilian housewives; thus, as historian Bryan McCann observes, it actually blended old and new elements of national culture.[73]

Although Gonzaga himself was from a relatively well-watered town in the interior of the state of Pernambuco (and thus did not have to migrate during droughts), he sang evocatively of the pain of migration and the deep longing to return home that shaped many sertanejos' lives. Gonzaga's first radio hit, "Asa Branca" (White wing)—still one of his best-known songs, based on a traditional melody—is a lament by a man forced to depart his land and love after his fields bake dry and his livestock die of hunger. Pleading with God to end the region's suffering, the man asks his beloved to remain faithful in his absence. The final stanzas describe the torment of living far from home and the wandering sertanejo's dream of returning to the sertão:

| | |
|---|---|
| *Hoje longe muitas léguas* | Now many leagues away |
| *Numa triste solidão* | In sorrowful isolation |
| *Espero a chuva cair de novo* | I hope for the rain to fall again |
| *Pra mim voltar pro meu sertão* | So I can return to my sertão. |
| | |
| *Quando o verde dos teus olhos* | When you cast your green eyes |
| *Se espalhar na plantação* | Across the fields |
| *Eu te asseguro, não chore não, viu* | I assure you, please don't cry |
| *Eu voltarei, viu, meu coração* | I promise to return, my heart.[74] |

A few years later, Gonzaga and Zé Dantas introduced a hopeful conclusion to this story. In "The Return of Asa Branca" the land becomes fertile once again and the once-tormented subject declares that he will marry his love, Rosinha, at the end of the year if the harvest is good.[75] Many of Gonzaga's songs describe birds as harbingers of drought or of the rain's return; birds were symbolically significant in the region because they migrate and thus could escape droughts relatively easily, whereas human families struggled to reach distant oases.

The songs written by Gonzaga and his contemporaries evoke the beauty of an abandoned homeland referred to fondly as the *terra natal* (native land), *berço* (cradle), *lá* or *lar* (hearth), *meu lugar* (my place), and an earthly paradise. These nostalgic sentiments spoke both to refugees from sertão drought and to other rural Brazilians who crowded into urban centers from the time of the Estado Novo onward, lured by the promise of expanding industrial labor. The idealized sertão became a metaphor for the more traditional society these migrants had abandoned.[76] One midcentury folk verse describes the sertão as an enchanted place where one must pass some time in order to have truly lived. Poet João Martins de Atahyde had moved (probably during a drought) to Recife, the capital of Pernambuco. His nostalgic poem portrays the lively world that drought destroys, describing the efflorescence following rain in the sertão and the energetic response by animals, farmers, and ranch hands enlivened by the earth's sudden fecundity and the promise of a good harvest.[77] Another patriotic cordel includes the refrain "Sou natural do sertão / . . . .conhecido por sertanejo" (I am native to the sertão / . . . .[a place] well-known to the sertanejo). Poet Pedro Alves da Silva dwells on the vigor and beauty of sertanejo life and the contentment he feels in his native surroundings, which provide everything he needs to live happily.[78]

Two songs recorded by Gonzaga during the mid-1950s explicitly address federal drought works. "Vozes de Seca" (1953) begins by thanking the elites (*doutores*) of the southeast for their aid to the sertão, then changes tone to chastise them for providing only charity to healthy men. This weakens sertanejo citizens, the lyrics assert, and makes them ashamed. In that climatically disastrous year, nearly half of Brazil lacked sufficient food, and Gonzaga warned that more lasting measures than mere alms were required to prevent further crisis:

| | |
|---|---|
| *Dê serviço a nosso povo* | Provide services to our people |
| *Encha os rios de barragem* | Fill the rivers by building dams |
| *Dê cumida a preço bão* | Give us food at lower prices |
| *Não esqueça a açudagem* | Don't forget the reservoirs |
| *Livre assim nóis da esmola* | Thus free from charity |
| *Qui no fim dessa estiagem* | At the end of this season |
| *Lhe pagamo inté os júru* | We will pay you, including interest |
| *Sem gastar nossa coragem* | Without sacrificing our honor |
| | |
| *Se o doutô fizer assim* | If you do this, sir, |
| *Salva o povo do Sertão* | Saving the people of the sertão |
| *Quando um dia a chuva vim* | When one day the rain comes |
| *Qui riqueza pra nação* | What wealth for the nation |
| *Nunca mais nois pensa em seca* | Never again will we dwell on drought |
| *Vai dá tudo neste chão* | All will move forward on this ground |

*Cumo vê, nosso destino*         As you can see, our destiny
*Mecê tem na vossa mão.*        Is in your hand.[79]

Here Gonzaga adopted the nationalist rhetoric of Vargas and regional boosters like Almeida, arguing that what was good for the sertanejo would bring prosperity to Brazil as a whole. But he also emphasized that what sertanejos desired most was independence: the capacity to support their families adequately on their own land without government assistance. Like the drought agency, and consistent with his own highly masculinized cowboy-bandit persona, Gonzaga viewed drought aid primarily from the perspective of men who could no longer survive as independent farmers.

In a song recorded two years later, Gonzaga praised the federal government for converting Bahia's Paulo Afonso waterfall into a hydroelectric dam. "Paulo Afonso" praises engineers for making Brazil's motto of "order and progress" a reality, saving its people from poverty through industrialization. The singer "hears the generator, happy messenger / saying with the force of the waterfall / 'Go forward, Brazil, go Brazil / go, go, go!' "[80] This song was probably a propaganda piece, perhaps commissioned for the power plant's opening (although that event did garner considerable attention). In this depiction, the fate of the northeast is again closely tied to national economic progress, echoing the integrationalist modernizing agenda of Vargas and his successors—particularly President Juscelino Kubitschek, who inaugurated the power plant. The lyrics extol several political figures by name, including Marechal Eurico Gaspar Dutra (president from 1946 to 1951, between the Estado Novo and Vargas's final presidency) and Vargas himself. Tragically, Vargas had killed himself a year earlier in the presidential palace, leaving a note to his supporters that blamed enemies of his populist agenda for his metaphorical crucifixion.

Popular northeastern songs and poems from the middle decades of the twentieth century do not present a uniform view of the sertão's needs or the state's responsibility to meet them. In some cases, public authorities are deemed negligent; in others, sertanejos are blamed for their own troubles. A number of poets pointed to poor government administration, favoritism in aid distribution, and neglect of sertãnejos' misfortunes as the root causes of misery during droughts. Others argued that even more ambitious projects of the sort already underway, and cooperation with those by the general population, were necessary to solve the sertão's recurrent crisis. Some verses may have been commissioned by politicians wishing to sway the views of their electorate. The varied discourse probably reflects a genuine spectrum of opinion among ordinary sertanejos about how drought should be understood and addressed, in light of their own circumstances and experience.

## The Limits of Technocratic Reform

In the American West, neither dry nor irrigated farming produced a landscape of family farms, as proponents of those technologies intended. Instead the federal government's plans became harnessed to powerful state and local interests that disregarded the original ideological motivations for reclamation projects. As several historians have argued, the technical efforts undertaken did not engender a democratic social order.[81] One reason for this was the relative impotence of agronomists when their recommendations conflicted with the desires of more politically influential sectors of the western U.S. economy. Similarly, in northeast Brazil, IFOCS's leading agronomists offered poor sertanejo farmers modest assistance to avert the most calamitous aspects of drought, but could not ensure equitable access to land and water. The reformist technocrats who staffed Vargas's drought bureaucracy from 1930 to 1945 and again in the 1950s achieved some economic and technological modernization, but did not dramatically diminish the vulnerability of many sertanejos to failed harvests. Like administrators of the U.S. Tennessee Valley Authority during the same period who operated within a landscape of racial segregation, Brazil's drought agency engineers and agronomists adapted their priorities to existing social and economic arrangements in the sertão. "In the end," Frederico de Castro Neves contends, "[Vargas's] New State [Estado Novo] wasn't so new after all."[82]

As heads of the drought agency's agricultural service during its first several decades, Trinidade and Duque approached sertão modernization in a manner consistent with Vargas's moderate progressivism. They hoped to incorporate sertanejos into their regional development efforts by introducing new technologies and encouraging more efficient production. The region's impoverished farmers were thought by many middle-class reformers to need the guidance of scientifically trained men, yet many sertanejos did not wish to adopt agronomists' cultivation methods. The technocrats' ambitions foundered on the shoals of humble sertanejos' distaste for intensive, irrigated farming as well as elite resistance to land redistribution.

Agronomists who today manage DNOCS's extensive irrigation zones told me in informal conversation that many sertanejo cultivators, when provided with government subsidized smallholdings of four to eight irrigated hectares, demonstrate little interest in adapting their farming practices and household economies to intensive farming. They prefer to grow subsistence crops as a source of household independence rather than planting potentially more profitable crops that leave them at the mercy of demand cycles and require loans to pay for technical inputs (repayable when the harvest is sold).[83] Across the twentieth century, Brazil's drought technocrats imported a middle-class, positivist vision of social progress that both elite and subaltern sertanejos were often wary of—though for quite different reasons. Well-meaning agronomists found no easily navigable "middle road" to regional development in the sertão.

6   Modernizing a Region

Economists as Development Experts, 1948–1964

· · · · · · · · · · · · · · · · · · · · · · · · · · · · · · · · · · · · · · ·

The 1950s witnessed a rigorous rethinking of rural development in north-eastern Brazil, guided by new federal agencies. These agencies were directed by economists rather than engineers. Two droughts at the beginning and end of the decade revealed that the efforts of the Departamento Nacional de Obras Contra as Secas (DNOCS; National Department for Works to Combat Droughts) over the prior half century had done little to reduce the human tragedy caused by the climatic scourge. This discredited the capacity of engineers to oversee regional development. In the midst of growing accusations that the federal drought agency had failed, economists were well positioned to assert their own expertise as regional planners.

The U.S. government had been among the first to define a significant role for economists as national and regional planners. American economists found employment from the progressive era onward in agencies like the Bureau of Agricultural Economics of the U.S. Department of Agriculture. Economists' influence over U.S. policy making expanded during the 1930s when they held prominent posts in President Franklin Delano Roosevelt's New Deal administration, overseeing the Tennessee Valley Authority (TVA) and other development organizations.[1] Brazil's first center for economic training and research, the Getúlio Vargas Foundation in Rio de Janeiro, was established in 1944. It was quickly followed by economics programs at Brazilian universities. After World War II, a number of young Brazilians received United Nations fellowships to study economics abroad. This new technocratic cohort challenged engineers' preeminence as Brazil's development specialists.

In the northeast, economists offered an approach to regional planning that looked beyond drought and its aftermath. In discussing the "northern problem" (*problema do norte*), economists distinguished between climatic instability and the social crises that droughts precipitated. Led by Celso Furtado, a Nordestino well versed in his region's colonial history, the new cohort of planners came to view drought and famine as *symptoms* of the northeast's imbalanced social and economic organization. They emphasized linkages between production in the semiarid sertão and on the coasts, and they proposed changes to employment patterns as well as to physical infrastructure. Their analyses were influenced by new international models, including those emerging from the TVA, and by foreign advisers familiar with those approaches.

Furtado received his training in development theory from Argentine economist Raul Prebisch at the UN's Economic Commission for Latin America. Adopting Prebisch's center/periphery model of relations among global regions, Furtado criticized fiscal policies that had contributed to disparities between Brazil's stagnating Nordeste (Northeast) and its industrializing southern states. He outlined a two-pronged strategy of industrial investment and greater food production that he believed would make northeasterners more resilient to drought and other misfortunes. As first director of the Superintendência de Desenvolvimento do Nordeste (SUDENE; Superintendency for Northeast Development), created in 1959, Furtado defined regional underdevelopment as the root cause of the drought problem. Instead of focusing on climate and hydrology, as DNOCS had done for fifty years with limited impact on drought victims, Furtado and his staff examined the larger social context in which droughts occurred.

SUDENE was authorized to set policy for the entire northeast and review proposals from other agencies, including DNOCS. Yet Furtado encountered numerous political obstacles to restructuring the region's economy. The most controversial of his development schemes was an irrigation law intended to expand food crop production as insurance against famine. The law would have required expropriation of certain categories of sertão properties to form smallholder colonies. Landowners associated it with another reform movement that overtly jeopardized their privileged position within the region. During the progressive years of the mid-1950s, rural workers began agitating for legal rights similar to those that had been granted to urban workers two decades before. Some farm laborers organized groups called Peasant Leagues and managed to obtain title to the land they worked. In this context Furtado's plan to reorganize the northeast's agricultural economy threatened conservative sectors of Brazilian society. In 1964, members of the middle and upper classes supported a military coup that led to the exile of Furtado, his closest colleagues, and other left-leaning reformers across the country. Under military rule in the 1960s and 1970s, SUDENE resumed its focus on northeastern industrialization absent the goal of providing greater food security to the poor.

Economists were optimistic about the potential transformative impact of foreign development models on the northeast. Yet the technical aid that they proffered was applied in a political environment that offered limited possibility for social transformation—certainly not on the scale that Furtado and his allies envisioned. Both land redistribution and economic reorganization were strongly resisted by elites—particularly northeastern estate owners and southern industrialists—who exerted disproportionate influence in the national government.[2] Furtado was unusual among twentieth-century sertão reformers for his ability to combine technical expertise with considered analysis of the northeast's history and social organization, yet even he underestimated the level of opposition to change that conservative segments of Brazilian society could muster when threatened. As in the early 1930s, the outcome of Brazil's brief period of progressive momentum from 1955 to 1964 was reactionary authoritarianism that undid much of SUDENE's early achievements.

## The Decline of DNOCS's Credibility and the Growth of Economic Expertise

By the time of President Getúlio Vargas's suicide in 1954, the costs to rural areas of his focus on urban industrialization were clear. Academics and government planners agreed that rural areas had been left out of the prior decades' economic expansion. This was detrimental to the national economy, since urban workers depended on rural producers for their food supply. Demographers monitoring national population growth recommended establishing colonies of small farmers throughout Brazil to increase food cultivation. These settlements would have to be provided with medical services and modern infrastructure or farmers would continue fleeing the countryside, chasing the mirage of greater opportunity in already crowded urban centers.

From 1932 to 1951, when no severe drought occurred to trigger mass exodus, the sertão's population exploded. The 1950 census counted 8.5 million sertanejos, nearly three times the estimated figures following the early 1930s drought.[3] By 1957, DNOCS reported the sertão population as 12.6 million. This apparent increase of 50 percent in seven years presumably reflects variations in how sertão territory was demarcated by different federal agencies. (The "drought polygon" sometimes referenced by DNOCS to designate areas affected by drought extends beyond the semiarid climate zone.[4]) Nevertheless, over two drought-free decades the number of inhabitants had clearly grown substantially, aided by continued reduction in infectious disease incidence and a cultural preference for large families, reinforced by Catholic teaching. This demographic trend meant that even minor droughts would affect an increasing number of people for whom food, refuge, and medical care had to be supplied rapidly, placing a heavy burden on the federal treasury.

Fueled by this need to increase the food security and economic productivity of sertanejos, many public figures opposed to the northeast's entrenched oligarchy expressed dissatisfaction with DNOCS's accomplishments. The agency's progress in irrigating sertão farmland was particularly unimpressive. Although drought inspector Luis Vieira claimed to have ten thousand hectares irrigated when his term ended in 1943, a report by his successor in 1950 stated that only two thousand hectares were irrigated; Vinícus Berredo explained that difficult negotiations with the owners of land surrounding DNOCS's reservoirs had slowed the construction of irrigation networks.[5] By 1958 the irrigated area around public reservoirs had expanded to six thousand hectares. This amounted to only 4 percent of the area that agronomist José Guimarães Duque deemed to be irrigable at that time—and less than 1 percent of the 800,000 hectares that Duque believed could be irrigated if the agency made that its top priority.[6]

As a step toward reorienting northeastern development priorities, President Eurico Gaspar Dutra (who held office from 1946 to 1951, in the interim between Vargas administrations) had experimented with "integrated river basin development" of the São Francisco River valley. He created the Comissão do Vale do São Francisco (CVSF; São Francisco Valley Commission) in 1948, modeled on the Tennessee Valley Authority. The TVA had been formed in 1930 when economic opportunities in the U.S. Southeast shriveled, a result of the drastic drop in crop prices and widespread unemployment that accompanied the Great Depression. Administrators for the TVA aimed to foster small-scale industry and improve agricultural productivity to spur regional economic growth. Their efforts received an unanticipated boost from the rise in manufacturing demand during World War II, which the valley's hydroelectric plants and nascent industries were poised to accommodate.

Brazil's CVSF was charged with coordinating development of trade, agriculture, and industry along the northeast's only significant perennial waterway and exploiting the river's potential to generate hydroelectric power. The CVSF was an autarchy directly subordinate to the president, not under the authority of any ministry; it was allotted one quarter of the

4 percent of federal tax revenues that had been allocated in Brazil's 1946 constitution for northeastern development and drought relief. In 1950 the commission contracted with the Ministry of Education and Health to survey health conditions in the valley and recommend sanitation improvements, in cooperation with the Serviço Especial de Saúde Pública (Special Public Health Service) funded by the Institute of Inter-American Affairs. Their summary report paints a bleak picture of rampant disease, poor nutrition, and minimal medical services, and emphasizes the need to educate the population about the importance of hygiene to improve life expectancy and productivity. The authors note that residents typically believed clear water to be clean and healthful, and were therefore not interested in filtration systems.[7]

Although the CVSF's initial plans were tremendously ambitious, including irrigating sixty thousand hectares along the river's banks, it ended up having little impact. Vargas devolved authority over the São Francisco River's largest waterfalls, Paulo Affonso, to a separate entity, the Companhia Hidrelétrica de São Francisco (CHESF; São Francisco Hydroelectric Company), whose formation he authorized in 1945 at the urging of Minister of Agriculture Apolônio Sales. Sales viewed the waterfall site as ideal for establishing agricultural colonies, particularly if the falls' potential for electric power generation were fully realized. As a mixed public-private corporation required to obtain much of its funding from investors, the CHESF was more efficiently administered than its fully public counterpart. The hydroelectric company began with the goal of doubling the northeast's electrical capacity and soon decided to triple it. By the late 1950s, all of this electricity was in use, and the CHESF undertook construction of a second power plant with twice the capacity of its first.[8]

Without authority over the portion of the São Francisco River most valuable to northeastern industry, the CVSF floundered in a sea of small and poorly administered projects. It proved ineffective in providing new opportunities along the river valley, and DNOCS accomplished little during the 1930s and 1940s to reduce sertanejo farmers' vulnerability to harvest failure; thus, the drought that struck in 1951–52 produced grimly predictable scenes of desperate families migrating to overcrowded refugee camps. At the camps, half-starved men registered to work on DNOCS projects in return for basic food for their families. The agency borrowed money from private lenders to supply these provisions, at additional cost to the public.

This lamentable yet familiar drama led to more vigorous criticism of DNOCS. As the agency's failings became starkly apparent, prominent journalists took up the call for a new approach to northeastern development.[9]

One military officer from the northeastern state of Alagoas asserted that the armed forces should take over regional development from DNOCS, using drought relief operations as a form of training for young soldiers—a "practice war," with the climatic scourge as their enemy.[10] Proposals to relocate some sertanejos to fertile regions just outside of the drought zone as an economical alternative to irrigation became commonplace.

In a speech to Brazil's legislature, José Américo Almeida, the minístro de viação e obras públicas (MVOP; minister of transportation and public works), who had been reappointed by Vargas in 1953 to oversee drought aid, testified that many of DNOCS's reservoirs had proven of little use when the drought struck. He suggested that multiple small reservoirs, well sited to enable irrigated farming, would be more useful than the large ones to which DNOCS had devoted most of its resources.[11] Lack of food reserves was the sertão's most pressing problem, but the drought agency ceased to focus on it once normal harvests returned. A nutritionist who toured the drought zone in 1953 reported on the meager sustenance provided to migrants employed at federal work fronts: 1,026–2,576 calories per day, mainly of carbohydrates, for men working long hours in scorching heat and already weakened by prolonged undernourishment. Like Almeida, Antônio da Silva Mello believed that the goal of reservoir construction should be increasing food and fish harvests. He concurred with many agronomists that only small farmers with secure land rights, and provided with instruction in new cultivation techniques, would make irrigated farming a success.[12]

## Vargas Establishes the Banco do Nordeste do Brasil

In an effort to rethink aid to the drought zone, President Vargas established the Banco do Nordeste do Brasil (BNB; Bank of Northeast Brazil) in 1952, staffed primarily by economists. That same year he established the Banco Nacional de Desenvolvimento Econômico (BNDE; National Economic Development Bank), cementing his administration's commitment to development planning grounded in economic analysis. Both institutions were intended to foster industrialization, which had fueled economic growth in many Latin American countries since the Great Depression (thanks to import-substitution policies pursued by Vargas and other national leaders). The BNB was established separately from the national bank to guarantee that a portion of development resources would be targeted at that region, since the southern region's more rapid industrialization lured both investments and trained staff there. The BNB was a publicly and privately funded entity like the successful CHESF; it was expected to assess the northeast's economic potential and loan money to promising projects that would eventually attract private capital, thereby freeing government funds for other efforts.

The BNB's federal funding came from the same revenue reserved for northeastern development that DNOCS depended on, and both organizations' mandate centered on adopting preventative measures to avert drought crises; thus, the bank's establishment was an acknowledgment by Vargas that the original northeastern development agency had not succeeded. As the president explained during his annual address in 1951, the BNB would reorient regional development away from DNOCS's focus on hydrologic infrastructure: "The very name 'Works Against the Droughts' conveys [the drought agency's] limitations, since it focuses the problem chiefly on engineering projects. It is timely, in view of past experience with modern regional planning techniques, to adopt a definitely economic and social approach for the study and solution of the [drought] problem."[13] Stefan Robock, an American economist hired by the UN to help the BNB during its initial years, later identified the bank as marking "the first official government acceptance of the economic solution [versus an engineering one] as federal policy for the Northeast. However," Robock cautioned, "the new policy was adopted as a supplement to rather than a substitute for the hydrologic solution [pursued by DNOCS]."[14]

In 1953 the BNB presented a plan for sertão development that enumerated the weaknesses of Brazil's existing approach to drought alleviation. These criticisms echoed the complaints that had been made about DNOCS's operations, including those from its own staff, since the agency's early years. There were insufficient scientific studies of the sertão on which development projects could be based. Federal support for the drought agency was inconsistent and thus could not be relied upon from year to year. DNOCS's work was not coordinated with that of related agencies like the Ministry of Agriculture, whose Directorate of Lands and Colonization was establishing experimental agricultural settlements on the borders of the drought zone. Farmers were not well versed in modern and ecologically sound methods, so they required training to adopt new practices without damaging their land.

The BNB's primary concern was the lack of an overarching plan to guide the efforts of all federal entities working on sertão development: "We lack a general program that would comprise in one organic whole an approach to the region's demographic issues; a plan for public works; economic organization, principally of agriculture and industry, designed in harmony with the imperative of conserving natural resources; a coordinated system among federal agencies in the region; and adequate planning for all of them, including financial entities, all working cooperatively with other public agencies and private initiatives."[15] In reviewing existing efforts to aid the northeast, the BNB's director noted the plethora of organizations involved; whatever problems persisted in the region, "[they] are not due . . . to a lack of institutions, funding, and federal legislation." The fault lay with the insular, disconnected work of the existing agencies. By 1956 the BNB had adopted coordination among regional development programs as its particular responsibility "in order to formulate a unified ideology of development, an indispensable step toward intensifying the pace of planning."[16]

The bank's analysis of the sertão's predicament looked beyond the narrow hydrologic definition of the drought problem that had long been accepted by DNOCS. This wider-ranging approach to regional planning was closely modeled on the TVA's work. When Vargas established the bank, the United Nations sent development economist Hans Singer on a three-month mission to evaluate northeastern Brazil's development potential. Based on Singer's favorable impression, the UN assigned Stefan Robock to work with the BNB for two years, from 1954 to 1956. Robock had been the chief economist for the TVA, and in that position he encouraged the simultaneous expansion of agricultural production and small-scale industry in the U.S. Southeast with the aim of increasing per capita income.[17] He was attentive to the social organization underlying the region's economy—his own surveys for the TVA included an analysis of African Americans' position in the southeast's economy.[18]

Robock reported back to the UN that planning for northeastern Brazil suffered from a severe shortage of trained technical personnel and reliable data. As a result, he changed the goal of the UN mission from "preparing a plan" to "creating a continuing planning process."[19] Robock established the Escritório Técnico de Estudos Econômicos do Nordeste (ETENE; Technical Office for Economic Studies of the Northeast) within the BNB and, aided by UN advisers, launched a yearlong program of economic analyses and other surveys. This project had the dual purpose of helping to establish a sound empirical base for regional development while providing valuable research experience for the Brazilian staff involved.

ETENE's first report described northeastern Brazil as the third poorest "country" in the world, ahead of only Burma and India. The average per capita income was less than half that of Brazil overall, and productivity per hectare in the agricultural sector was extremely low. Nordestino farmers reaped seventy kilograms of cotton per hectare, compared with 104 in India, 214 in the state of São Paulo, and 304 in the United States. The average sugarcane yield was thirty-eight kilos per hectare, compared with forty-eight in São Paulo and sixty-nine in Puerto Rico.[20]

ETENE's numerous subsequent reports were written by Brazilian and American economists (the latter sponsored by the U.S. State Department's Point IV Mission) and technical advisers provided by the UN's Food and Agriculture Organization. These analyzed reasons for the low productivity of northeastern farmers and suggested alternate crops and farming techniques that would produce higher yields. Other publications traced population growth trends and offered strategies to promote migration from the northeast in ways that would benefit the regional and national economy. Researchers surveyed the basic social development needs of the northeast, including better health care, sanitation, and transportation. They estimated an illiteracy rate of 74 percent and an infant mortality rate of over four hundred per thousand births in some areas, more than three times the rate in São Paulo. Yet there were few trained personnel in the region to address such issues.[21]

Reflecting on the outcome of his work in Brazil, Robock identified ETENE's studies as having provided empirical support for a significant shift in development thinking that the BNB's staff had begun to articulate when the bank was established. During the 1950s, regional planners came to believe that "the most basic problem of the Northeast is not the periodic drought but continuing poverty."[22] In Robock's understanding, that poverty was the result of poor soils, insufficient industry, slow adoption of modern technologies, underemployment, paltry investment in human resources, and a demographic imbalance toward the young—most of these being issues that previous drought aid had ignored. Robock's recommendations for the northeast mirrored those of Brazilian development economists like Celso Furtado through the early 1960s.

Robock's assessment of the complex of factors that turned the northeast's climate fluctuations into humanitarian disasters was markedly different from DNOCS's focus on transportation and hydrologic infrastructure. This broader perspective led him to recommend a variety of development measures, many only indirectly related to drought. In addition to reorganizing agricultural production and encouraging industrialization, Robock believed it essential that the government invest in basic education—particularly literacy. This would make new technologies easier to introduce, he imagined, and it would also expand the electorate, since the right to vote still carried a literacy requirement—thus encouraging politicians to be more responsive to the most marginal Nordestinos. With regard to population growth, Robock proposed that either birth control be encouraged (though this was difficult to promote in the strongly Catholic country) or that the federal government fund schools and health care on the grounds that surplus workers from the northeast formed an important labor pool throughout Brazil.

### President Kubitschek's Vision for Northeast Development

After Getúlio Vargas's shocking suicide, Juscelino Kubitschek and João Goulart were elected president and vice president, respectively, in 1955. Kubitschek and Goulart were Vargas's political heirs, and they continued—even amplified—his industrialization platform, supported by urbanites of all classes. President Kubitschek directed his administration "to the central purpose of getting the most possible industrialization and development in the five short years which he had at his disposal," a goal summarized in the slogan "fifty years in five!"[23] In pursuit of this agenda Brazil's state-subsidized industrial capacity increased beyond market demand for its manufactured products, making federal finances precarious—and leading to conflict with the International Monetary Fund (IMF) as a result. Kubitschek identified energy production and transportation as bottlenecks that needed to be addressed before other development plans could move forward. Given his ambitious goals, it was essential that he retain the support of Brazil's wealthiest and most populous region. He focused primarily on the southeast, arguing that its large market and modern infrastructure made industrialization easier to implement there than elsewhere. In 1956, 83 percent of national revenue was generated in the south, as compared with 13 percent in the northeast.[24]

Nonetheless, Kubitschek provided significant aid to the northeast, increasing the BNB's budget fivefold and allowing DNOCS to substantially expand road and dam construction. He encouraged the agency to complete the Orós Dam, its biggest reservoir project, which had languished for three decades. The president established the Departamento Nacional de Endemias Rurais (National Department of Rural Diseases) in 1956, combining the malaria, plague, and yellow fever services previously contained within the Departamento Nacional de Saúde (National Health Department). He hoped to focus more aggressively on rural reforms during a second presidential term but was never given that chance. Brazil's 1946 constitution stipulated that presidents could not serve consecutive terms, which barred him from the 1960 elections, and the military staged a coup before the 1964 elections (in which he could have run) were held.

In May 1956 the northeast's Catholic bishops convened to discuss the region's economic problems and demand greater government commitment to reducing hunger and poverty. Rio de Janeiro's auxiliary archbishop, Dom Helder Câmara (a Nordestino who would become Archbishop of Recife/ Olinda in 1964) organized the meeting, and Kubitschek delivered an address sympathetic to the prelates' concerns. In December of that year the president formed the Grupo de Trabalho para o Desenvolvimento do Nordeste (GTDN; Working Group for Northeast Development) to investigate the BNB's assertion that regional development efforts had to be better coordinated. The GTDN, led by Celso Furtado, included representatives from the BNB, the National Development Bank, the Ministry of Transportation and Public Works, the CVSF, and the Ministries of Agriculture, Health, and Education. Kubitschek asked the group to meet regularly for a period of two years and present him with a regional development proposal. The GTDN's report, submitted in 1959, became the basis of northeastern development strategy over the following five years, and it reiterated many of Stefan Robock's earlier recommendations.

The GTDN's report explained drought crises as resulting from a dangerous combination of overpopulation and underproduction of food. It argued that DNOCS had not adequately increased the cultivation of subsistence crops in the northeast, which was why sertanejos could not survive droughts without federal assistance. The water stored by the drought works agency had sustained cattle more than agriculture, since few reservoirs were linked to functional irrigation networks. DNOCS had arguably enabled overpopulation by employing excess sertanejo workers and regulating food prices during droughts so that too few laborers were forced to leave the area. "There is no escaping the conclusion that each and every measure which manages to augment the demographic burden without also improving the stability of the food supply contributes, in the end, to making the economy more vulnerable to droughts," the GTDN's members concluded.[25]

In highlighting inadequate food production as the central cause of the drought crisis, the GTDN reiterated an argument made starkly by Josué de Castro, a veteran campaigner against hunger, when he was serving as Pernambucan representative to the national legislature: "Famine in the Northeast . . . is a social rather than a natural phenomenon," he informed his colleagues. "Latifundism is the Siamese twin of technical obsolescence. The big estates pursue a primitive kind of agriculture, a proto-agriculture, without technical know-how, without fertilizers, without selective management, without mechanization. Everything is done in a most rudimentary way, draining the poor sertanejo in order to produce what turns out to be less than enough to satisfy his hunger."[26] In Castro's view, concentrated land ownership in the northeast contributed to the limited focus on food production, making droughts especially threatening for those with no reserves to draw on. DNOCS's agronomists had been arguing similarly for over a decade that irrigated smallholding was essential for regional food security. A few years later, Castro would introduce the first national legislation aimed at forcing more equitable land distribution in the public interest.

The GTDN proposed creating new jobs for sertanejo farm workers in industries that would be less affected by droughts. It recommended that the government invest in iron and steel works, which could form the basis of additional regional industries, and also encouraged the expansion of industries for which the northeast possessed primary materials like minerals and fibers crops. If more northeasterners were paid in cash for work outside of the agricultural sector, the group theorized, they would be able to buy food produced by the remaining farmers, bolstering the market for subsistence crops. The GTDN recommended subsidizing the migration of some sertanejos to the São Francisco River valley (which would have to be irrigated as part of this resettlement plan) and the western frontier of the drought zone, where the climate was more humid. Essentially the group advised intensifying agricultural and industrial production where that was possible and relocating some sertanejos outside of the drought zone to provide reliable food harvests for those left within it.

In summarizing its regional development strategy, the GTDN report stated,

> The first step is to make the Northeast more resistant to droughts.
> To do this, it is necessary to reorganize the economy of the semiarid
> zone, in ways different from what is currently being done. This step
> can only be achieved by . . . incorporating the economy of the
> unoccupied or semioccupied land of the Maranhão and Goiana
> hinterland [on the western border] into the region, and by more
> intensive use of the region's humid valleys. With the territorial
> expansion toward the humid lands of Maranhão, the disequilibrium
> between land and labor that characterizes the Northeast economy of
> today will diminish. The second step consists of raising the median
> productivity of the workforce concentrated along the humid

[coastal] strip, which necessarily requires intensification of industrial investments there. . . . Taken together, these two steps comprise a policy that . . . should be able to modify the region's economic structure.[27]

Here the GTDN extended consideration of the northeast's deficiencies well beyond the drought zone to encompass more fertile regions along the sertão's borders. The goal was a significant reorganization of the regional economy, increasing production in several sectors to support a growing population even during drought years.

### The Drought of 1958

In the midst of the GTDN's discussions about how to alter northeastern development strategy, another severe drought struck. Although it was less geographically extensive than the drought of 1932, which had affected three million people, roughly eleven million people were impacted due to population growth and the specific areas where this drought was most severe. DNOCS offered sertanejos voluntary resettlement in the Amazon and points south as one aid option. MVOP Lucio Meira reported in July that twenty thousand drought refugees had accepted the offer of relocation; he reiterated to the national legislature that a colonization program for points west of the drought zone should be included in regional development plans, as the GTDN would also propose.[28] (Meira, a personal friend of Kubitschek, became head of the National Development Council a year later and then directed the BNDE.)

Even more than during previous droughts, accusations of corruption in the distribution of aid abounded. Among the reported abuses were falsified papers drawn up by political bosses in order to receive payment on behalf of imaginary work crews and excessive charges by local *fornecedores* who provided food to the migrants and recouped expenses (plus interest) once funds arrived.[29] A physician who began working for DNOCS in 1958 reported later on the devastating conditions prevailing in Fortaleza, Ceará, where twelve to fourteen thousand migrants were crowded into an encampment without suitable sanitary arrangements. Typhus and diarrhea outbreaks ensued, and relatives of the sick were asked to bury their diseased family members' excrement to prevent further contamination. Doctors in the camp's hospital were accompanied by guards who sprayed DDT to eliminate flies while they attended patients. This physician commented that the "macabre" role played by *fornecedores* was not entirely DNOCS's fault, since the agency was simply not equipped to deal with an emergency on that scale and had to improvise in order to provide rapid assistance.[30]

Josué de Castro led a presidential commission to evaluate the work of federal development agencies during the drought, and in his report to Kubitschek and the legislature he was particularly critical of the BNB. During this first rigorous test, it became clear that it was unprepared to launch new projects on behalf of drought victims.[31] In response to questions from journalists, Castro reiterated that the prevalence of large estates in the rural northeast was the ultimate cause of the drought crisis, because this led to inefficient use of land and, consequently, low agricultural output.[32]

Following the drought, ETENE documented the scant impact of government aid on the affected population and proposed ways to increase the resilience of ranching and other key economic sectors. The accumulating evidence that DNOCS and other federal agencies had failed to fulfill their responsibilities in the drought zone made cries for change more urgent. Kubitschek requested yet another report on the region, this time from Furtado, now a director of the BNDE.

Furtado's credentials for elaborating a regional development plan were impeccable. He was born in the interior of Paraíba state and raised in Recife, the region's largest capital. He attended law school (one of the main avenues to a political career) in Rio de Janeiro, then joined the military during World War II and served briefly in Italy. Following the war Furtado received a doctorate from the Sorbonne for his dissertation analyzing Brazil's colonial economy. Upon returning to his native country as minister of finance in 1948, he joined the newly formed Fundação Getúlio Vargas as a staff economist. Furtado was hired almost immediately by the UN's Economic Commission for Latin America (ECLA), headquartered in Santiago, Chile.

ECLA was the first institute for economic study in what Raul Prebisch, its executive secretary from 1949 to 1963, termed the "underdeveloped" world. Prebisch conceptualized the global economy according to a center/periphery model, in which a dependent agricultural sector supported the industrialization of wealthier regions. ECLA defined underdevelopment as the mixture of traditional economic activity with more modern elements. Its staff saw industrialization as the key to economic growth, since this had been the experience of Latin American countries since the 1930s (and of Europe and the United States nearly a century earlier).

Furtado quickly became the director of ECLA's Development Division, and in 1953 he relocated to Rio de Janeiro to head the BNDE as well. After five years Furtado left his position with ECLA permanently to remain with the BNDE; by that time he had founded the Clube dos Economistas, which published a journal (the *Revista Econômica Brasileira*) that was critical of free-market neoliberalism on nationalist grounds. President Kubitschek and Brazilian industrialists strongly supported Furtado's approach to national development.

With regard to the northeast, Furtado interpreted its problems in more historical terms than Robock, other foreign advisers, or DNOCS's managing engineers (most from other regions of Brazil) had done. Based on his dissertation research, he viewed the northeast as a colonial appendage of both Brazil and the global economy, and he saw this as the reason for its underdevelopment. Furtado's proposals for economic reform highlighted inequities in land distribution and political power that he believed contributed to the northeast's low productivity. In "A Policy for the Economic Development of the Northeast," submitted to Kubitschek in 1958, Furtado called regional inequities "the most serious problem to confront at the present stage of Brazil's economic development."[33] He cited several fiscal policies that had inadvertently retarded the northeast's growth while aiding the South. Regressive tax policies meant that wealthier regions contributed a smaller proportion of their income to the federal treasury than poorer ones. Federally subsidized industrialization directed disproportionate support to the south-central region, which was better able to expand industry than the northeast was. Taxes on exports, meant to compensate the government for importing the capital goods needed by new industries, hit the northeast hardest, since those states exported more than they imported. Such misguided measures had unjustly crippled the northeast in the service of the national economy, Furtado argued. When the federal government did contribute significantly to the region, as it did during droughts, the funds were largely used for temporary aid measures rather than for investments to expand opportunity over the longer term.

The remedy for these regional imbalances, Furtado maintained, was for the federal government to promote investment in northeastern industrialization. Government agencies should train a northeastern "managerial class with a flair for development."[34] Like the GTDN report that he had overseen, Furtado's policy paper contended that drought aid had unintentionally contributed to the northeast's instability by temporarily supporting a surplus population without making those people less vulnerable to the next calamity. He proposed reorganizing the northeast's agricultural economy to focus on providing food to the urban population, which would increase as new industries offered desirable jobs to underemployed sertanejos. With a re-distributed workforce and more emphasis on food cultivation, fewer Nor-destinos would go hungry when droughts struck. In Furtado's evolving understanding, drought and subsequent famine were symptoms of the northeast's precarious economic organization rather than the cause of its difficulties. Whereas DNOCS's managing engineers had placed drought at the center of the northeast's instability, Furtado relegated climate issues further into the background.

## CODENO and Agrarian Reform

After receiving Furtado's 1958 report, Kubitschek created the Conselho de Desenvolvimento do Nordeste (CODENO; Economic Development Council for the Northeast) in February 1959. The council, headed by Furtado, was a temporary body charged with drafting a plan for coordinating regional development efforts. Its members included the nine governors of northeastern states, representatives of federal ministries involved in the region, and the heads of several development agencies. In June 1959 Furtado presented CODENO's perspective on northeastern development to Brazilian military officers gathered in Rio de Janeiro.[35] As in his prior report to Kubitschek, he emphasized the need to remedy the exploitative relationship between Brazil's industrializing south-central region and the northeast, in which the latter provided raw materials for manufacturers abroad and in the wealthier region. During Brazil's First Republic, politicians from the increasingly influential south had arranged for special federal price supports to aid coffee producers, which strengthened their economy at the expense of the declining sugar zone.

Furtado explained the contrasting economic and social development of the northeast and south as stemming from their respective histories as sugar-producing versus coffee-producing regions; landholding patterns reflected the differing demands of sugar and coffee production. Sugar planters in the colonial northeast needed substantial capital to construct and maintain a mill, and they required large plantations in order to make their operations profitable (as did the ranchers who first settled the sertão to supply meat, hides, and draft animals to the sugar zone). This legacy remained evident through the twentieth century. Data collected for 1960 indicate that two-thirds of agricultural properties in the northeast occupied less than 5 percent of its land area (taken all together), while the largest 1 percent of estates occupied almost one-third of the region.[36] In contrast, coffee, which had flourished as an export crop in the states of Rio de Janeiro and São Paulo in the nineteenth century, could be profitably grown on small family farms. Its production required little start-up capital, and many producers continued to operate at a small scale into the twentieth century.

Labor arrangements formed another important contrast in the economic history of the northeast versus the south, according to Furtado. The northeast's sugar barons depended on slave labor from the sixteenth to the late nineteenth centuries. This hindered the development of an internal market for consumer goods, and the laborers emancipated in 1888 received very low wages, which slowed economic growth. Most southern coffee farms, on the other hand, were established after the Atlantic slave trade was abolished in 1850. The majority of their workers were European immigrants who expected cash payments and a higher standard of living than did rural northeasterners resigned to the slaveholding regime.

Furtado observed that even settlers in the northeastern interior were heavily dependent on the sugar economy. Ranchers provided meat, milk, and leather to coastal residents, and their cattle were the mills' power source. Small farmers in the *agreste*—the transitional region between the coast and sertão—were generally former plantation workers who left to try their luck in the hinterland when sugar exports slowed. Neither of these sectors had an economic momentum of its own; both followed the fortunes of the sugar trade. CODENO's goal was to identify new ways of employing Nordestinos so that the regional economy would be more resilient to both climate fluctuations and changes in export markets. In particular, the council wanted to make food crop harvests more reliable.

To this end, council members drafted an irrigation law in August 1959 with support from Kubitschek. This proved to be CODENO's most controversial proposal. DNOCS agronomist José Augusto Trinidade had supported a similar law in 1940, because he feared (presciently) that irrigation canals would enrich a few large landowners while the reservoirs that fed them displaced many poor people. Trinidade's bill garnered little political support, and CODENO's renewal of the debate over irrigated settlements encountered similar obstacles. To justify revisiting the question, CODENO pointed out that DNOCS's expenditures on irrigation had not yet diminished the food shortages caused by droughts, despite the efforts of Duque and others to make good use of water accumulated in the agency's dams.

CODENO's irrigation bill proposed that all farms surrounding the drought agency's reservoirs be expropriated from their owners if they were greater than thirty hectares in size or not under cultivation. The same law would apply to farms along irrigable rivers, except that those were expropriated if they exceeded one hundred hectares. The council expected to recoup the cost of land expropriation through rents and taxes for water use charged to farmers who settled the new colonies. Landowners were to receive compensation for their property according to a price table, determined by CODENO, based on land and soil surveys that DNOCS had conducted. This compensation would not take into account increasing land values that resulted from existing or planned irrigation works. Passing the law therefore required an amendment to Brazil's 1946 constitution, which stipulated that land expropriated by the government had to be compensated at market value (which presumably rose following the installation of irrigation works).

In his June 1959 presentation to military officers, Furtado had advocated a cautious approach to land expropriation. In most of the sertão, low soil fertility required that farms be large in order to remain economically sustainable. The best option in those areas, he believed, was to mechanize agriculture and create alternate jobs for the laborers made redundant by this process. Yet in more fertile parts of the northeast, production could be intensified through irrigation; in those areas, underused land should be redistributed to small farmers.

Furtado adopted a modest tone when speaking to the officers, noting that land expropriation was ultimately a political matter and beyond his purview as a technical adviser. "What I cannot do is to cloak my technical qualifications with any political banner. Before doing that, I would need to explain that I am speaking simply as a citizen or novice politician," he demurred.[37] Furtado's obligation was merely to determine what options were economically feasible, after which politicians could decide which policies to pursue. Nevertheless, given the public expense entailed in providing irrigation, it seemed reasonable to expect that the investment would yield socially desirable ends, he contended—such as allowing year-round cultivation of food crops. DNOCS's irrigation works had enriched some individual landowners without noticeably benefiting the wider society, Furtado asserted. As to the question of how the government could legally control the use of private land, he suggested that federal agencies could legitimately manage the distribution of water into canals, making irrigation available only to farmers who used their land for socially desirable ends.

Furtado's careful articulation of the need for irrigated smallholding and the terms under which an irrigation law might be implemented won him the support of the military representative to CODENO.[38] But the council's proposed law was strongly opposed by the government of Ceará, one of the states hardest hit by droughts. The president of Ceará's agronomic society, speaking on behalf of the governor, argued that expropriating productive farms in the drought zone was senseless; Francisco Alves de Andrade claimed that many of Ceará's irrigated properties were quite small (under six hectares), but even the larger ones should be encouraged and provided with soil conservation assistance, to create a solid base for regional development. If land expropriation were undertaken in the northeast, Andrade asserted, it should start along the sugar-growing coast (and thus not in Ceará); that area had a much greater proportion of large estates and was more easily farmed. He contended that growing support for agrarian reform was an attempt by urban industrialists to deflect attention from legislative efforts to increase their workers' salaries—a more useful measure to spur regional development. Hasty and ill-considered reform could leave Ceará in worse condition, Andrade warned, particularly if the fifteen-hectare plots that CODENO proposed distributing to farmers proved insufficient to sustain a household, even when irrigated. He asserted that successful agrarian reform should be based on studies of sertão landholding and social organization, yet no such studies had been conducted.

Andrade maintained that small farmers would be better served by forming credit associations to help them invest in the land they already worked. The top priority for the sertão should be increasing production, as the TVA had done, not worrying about equitable distribution. It was futile to "socialize poverty," he cautioned.[39] The result of Andrade's vigorous criticism (presumably on behalf of influential *cearense* estate owners and ranchers) and other discussion among CODENO members was that the size of plots to be distributed to farmers was increased to twenty-five hectares around reservoirs and one hundred hectares along rivers, with existing landowners permitted to retain up to two such lots for themselves. Otherwise the bill presented to Kubitschek was largely unaltered.[40]

DNOCS also criticized CODENO's recommendations for Northeast development. The agency was clearly threatened by the formation of a new organization directly subordinate to the president and authorized to oversee northeastern development policy. DNOCS director José Cândido Castro Parente Pessoa wrote several memos in 1959 to MVOP Ernesto do Amaral Peixoto, complaining about Furtado's "defamation campaign" against DNOCS. (Peixoto was an engineer, a former *tenente*—one of the young military officers who attempted to overthrow the oligarchic federal government in the mid-1920s—and leader of one of the political parties founded by Vargas; he was also married to Vargas's daughter.) Parente Pessoa depicted Furtado as being ignorant of much of DNOCS's work, and he accused Furtado of inciting opposition to the drought agency—which, he warned, might lead to raids on food reserves and other rebellious acts that would endanger government personnel and cost substantial sums to redress. He asserted that Furtado had mobilized the press against DNOCS and demoralized the agency, which had always been "one of the strongest obstacles to communist and extremist infiltration in the interior Northeast."[41] Parente Pessoa insinuated that Furtado was himself a political extremist, most likely a communist.

Linking CODENO's leader to communism was a powerful accusation at that moment in Brazil. Many journalists already associated the council's proposed reforms with a growing movement demanding expanded rights for rural workers. In 1955, tenant farmers on a former sugar plantation in Pernambuco formed a cooperative that was responsible for paying the rent of all workers on the property. The absentee landowner, one Oscar Beltrão, was initially pleased with this arrangement because it eased his administration of the estate. But neighboring landowners convinced him that once his workers organized, they would make other demands. Beltrão tried to evict members of the new association, and they refused to leave unless he compensated them for the land they had farmed.

Beltrão's tenants sought legal assistance from Francisco Julião, a lawyer in Recife. Julião was a representative to the Pernambucan legislature, the only member from his state's small socialist party. A 1939 graduate of Recife's prestigious law school, and the grandson of wealthy sugar planters, Julião avidly read the work of prominent leftist intellectuals and came to adopt their ideology, based on his own experience of Brazilian politics. His efforts to establish the property rights of tenant farmers launched Brazil's agrarian reform movement. As more and more landowners dismissed workers during the 1950s in favor of mechanized production—and raised rents on their remaining tenants in order to pay for these investments—the formation of workers' cooperatives accelerated. Their rallying cry became the redistribution of land to sharecroppers and other farm laborers. Journalists termed these groups the Peasant Leagues, and many observers understood CODENO's irrigation law to be part of that movement.

Julião's organization of public demonstrations by rural workers, and his often inflammatory statements, made landowners and more conservative sectors of Brazilian society fear the Peasant Leagues. Castro would later remark that depictions of the Peasant Leagues as communist were exaggerated. "It is closer to the facts to see the Peasant Leagues as motivated by a primitive Christian spirit that still permeates the collective soul of the Northeast," he wrote from exile in 1966.[42] Any ideological overlap with Marxism in the rural workers' agenda was largely coincidental, Castro asserted.[43] He attributed the rise in rural activism during the 1950s to increased Nordestino contact with Brazilians from other regions, through migration and mass media. These interactions alerted poor northeasterners to the better economic circumstances and legal protections available elsewhere in the country.

## Promoting the Irrigation Law

Nordestino elites were resolutely opposed to any law that threatened landholding and labor arrangements, the sources of their political influence and personal wealth. When President Kubitschek submitted CODENO's proposed irrigation law to Brazil's legislature, opponents quickly drafted an alternate law of the same name. The revised law served landowners' interests in expanded irrigation without addressing the economic and social issues that concerned CODENO's members. Furtado realized that he would have to secure broad support for CODENO's agenda in the south-central region of Brazil in order to circumvent the conservative northeast voting block in the legislature. The months when CODENO's irrigation law was under debate were a harsh introduction to Brazilian political culture for Furtado. Opponents of CODENO's proposed law distributed a police report to all members of the legislature that depicted Furtado as a spy for the International Communist Party.[44] When Furtado obtained access to his dossier at the National Security Council, he found that it was full of unconfirmed allegations — many of which, he claimed, were completely fabricated. There were also minute details of his activities, all gathered while Brazil was experiencing one of its most democratic periods.

In order to draw national attention to the irrigation controversy and the shortcomings of existing federal aid for the northeast, Furtado sponsored a trip to the region in September 1959 by outspoken journalist Antônio Callado, who had reported sympathetically on the sertão's woes during the 1953 drought.[45] Following his second visit to the drought zone, made in the company of Furtado's brother (another CODENO member), Callado published a series of articles in Rio de Janeiro's *Correio da Manhã*. He hoped to help win support in the south-central region for making CODENO into a permanent agency that would coordinate the work of all northeastern development organizations. The articles, reprinted as a book in 1960, elicited considerable response from CODENO's supporters and detractors. In them Callado renewed his criticism of what he termed the "drought industry," a development sham with DNOCS at the center. "Rather than organizing to combat droughts, [northeastern] elites industrialized the drought," he wrote. "They live off it and the income it brings, not in spite of it. They needed a Bank of the Drought, which would nourish the calamity and its associated industries; and they found it in DNOCS."[46] DNOCS's colossal projects were so notorious in the northeast, according to Callado, that the word *federal* had been adopted in the local slang to mean "stupendous" or "enormous."

In reviewing DNOCS's efforts on behalf of sertanejos, Callado criticized engineers' dominance within the agency. He observed that Duque, an agronomist, had not been given a directorship despite twenty-eight years of admirable service. "The role of engineers in DNOCS should have been passed on ages ago," Callado asserted. "It is much more important now to make use of the already accumulated water in order to sustain subsistence agriculture."[47] Duque concurred with Furtado's analysis of the northeast's problems and admired his proposed solutions. He confirmed, in an interview with Callado, that water scarcity was now less of an issue than overpopulation and the lack of a coherent sertão development plan. Duque had become more optimistic and energized about regional change than he was during the 1940s, perhaps motivated by Kubitschek's ambitious vision for national development (epitomized by the rapid construction of Brasília as the CODENO debates were taking place).

Callado publicly accused drought inspector Parente Pessoa of using DNOCS funds to enrich himself and his relatives by routing a new road near his family's property and constructing reservoirs that were useful to them. Parente Pessoa defended himself in memos to the MVOP and letters to newspapers where Callado's columns appeared, insisting that no DNOCS projects were undertaken without good technical reason, and certainly none were selected by him out of personal interest.[48] As for apparent corruption in aid distribution during the 1958 drought, which Callado drew attention to, this was merely the result of mistakes made by the one hundred DNOCS staffers faced with 400,000 migrant men to employ and feed—a task that far exceeded their experience and expertise, Parente Pessoa contended.[49]

One aim of Callado's newspaper articles was to explain why CODENO's irrigation law was opposed by the state of Ceará, which suffered the most during droughts. Callado's analysis rested on the assertion that Ceará's economy depended on a form of slavery. The northeast's modern-day slaves were of every skin color—which made the arrangement even more shameful, in his view, than enslavement restricted to Africans and their descendants. Ceará's landowners, who operated outside the law in many respects, often paid their workers 40 percent of the regional minimum wage. The aid that DNOCS had made available increased their land values, secured their production, and sustained their cheap workforce through hard times. *Cearense* elites had no incentive to alter this system; they had no desire to give up part of their land in return for irrigation, and they did not want to lose their inexpensive workforce to irrigated smallholding.

Callado devoted several articles to descriptions of Francisco Julião's work on behalf of the peasant leagues; these articles were among the first to bring the agricultural workers' struggle to national attention and succeeded in garnering public sympathy for Julião's campaign. Under pressure to respond, the government of Pernambuco expropriated the Beltrão family's plantation in 1959, giving the land to those who had farmed it. This victory gave tremendous hope to the Peasant Leagues throughout the northeast, and the movement flourished.[50]

In the end CODENO's irrigation law did not pass. It had become too closely associated with Julião's agrarian reform movement, which scared conservatives as well as many moderates. Callado may have borne some responsibility for this, since he published inflammatory comments by the outspoken lawyer. For example, Julião asserted that his followers could easily burn down all of Pernambuco's *fazendas* (rural estates) within a few hours if they chose to; the fact that they had not done so was meant to indicate their commitment to operating within the law. But SUDENE, the permanent version of CODENO, did win legislative approval—thanks in part to broad public support for a new approach to northeastern development that Callado had helped to incite. Politicians from the south-central region voted to establish SUDENE over objections from DNOCS and some conservative Nordestino legislators. Once it was clear that SUDENE would be created, northeastern legislators hostile to Furtado tried to block him from becoming its director. Kubitschek pretended to consider this in order to ensure that the law creating SUDENE would pass. When the superintendency was finally established, in December 1959, Kubitschek appointed Furtado to lead it, over the objections of Furtado's foes.[51] The president hoped that SUDENE would solve the problems caused by uncoordinated northeastern aid efforts and lay the groundwork for future regional development.

In justifying the creation of SUDENE, Furtado argued that a new perspective must be adopted for northeastern development, since DNOCS had failed to mitigate the drought problem after fifty years—as the 1958 drought starkly confirmed. "The problem with the Northeast is not drought but rather regional underdevelopment," Furtado asserted. "There are areas with environmental conditions much more precarious than those of the Northeast whose populations enjoy relatively elevated standards of living."[52] This required a different kind of expertise from that of civil engineering or even agronomy, in Furtado's view—specifically, "specialists in economic development" (like himself).

The debate about SUDENE's formation took place at a time when Kubitschek was under pressure from the IMF to decrease government spending. The IMF cited high inflation and a poor balance of trade as signs that the president needed to reign in his ambitious expansion—particularly the construction of Brasília. Inflation was due in part to the government's subsidies for industry, which were rarely lifted once the industries matured. Kubitschek broke with the IMF in mid-1959, fearing that its stringent structural adjustment demands would provoke a recession. Furtado and many Brazilian nationalists applauded this decision and demonstrated their confidence in the president's agenda through their enthusiasm for SUDENE.

### SUDENE's First Guiding Plan

SUDENE's jurisdiction encompassed an area much larger than the "drought polygon," which distinguished it from DNOCS at the outset. It was to be a multifaceted regional development agency, not merely focused on drought. The area falling under SUDENE's purview stretched from the northeast coasts to the Amazon frontier and encompassed one-fourth of Brazil's sixty-six million people. The superintendency was established as an autarchy (like the CVSF), autonomous from federal ministries and subordinate only to the president. It was to receive 2 percent of federal tax revenues, separate from the 3 percent already shared by DNOCS and the BNB (of which the BNB received one-fifth) and the 1 percent allotted to the CVSF. Release of this funding was subject to legislative approval of the agency's Plano Diretor de Desenvolvimento Econômico e Social do Nordeste (Guiding Plan for the Economic and Social Development of the Northeast).

SUDENE's twenty-two-member Deliberative Council, appointed by the president, comprised (as CODENO's had) all the northeastern state governors along with representatives of federal agencies and ministries working in the region. Furtado hoped that this participatory arrangement would diminish hostility between SUDENE and other agencies with related mandates, fostering a more cooperative relationship. At times Furtado acceded to the demands of the northeast's most powerful governors in order to prevent their forming a bloc against his agency.[53] Participants on the Deliberative Council in its early years included renowned sociologist Gilberto Freyre (for the Ministry of Education), who had founded the Instituto Joaquim Nabuco in Recife a decade earlier to study the northeastern economy and culture. Duque represented the Ministry of Transportation and Public Works, responsible for DNOCS. SUDENE's members were charged with setting development policy for the northeast and reviewing proposals from other agencies that impinged on matters discussed in SUDENE's Guiding Plan.

SUDENE produced its first Guiding Plan three months after it was formed. The document's major emphases were industry to provide more jobs; roads to support industrialization; and reorganizing agriculture to supply food to urban workers. The superintendency planned to drain farmland in coastal valleys, irrigate parts of the sertão, and expand food crop cultivation along the frontiers of the drought zone. SUDENE hoped to stabilize the region's food market, which droughts often devastated, by warehousing emergency food stores. The Guiding Plan also called for improvements to the northeast's fishing and cattle industries to make more protein available to Nordestinos. Not surprisingly, many of these objectives followed the policy recommendations made by the GTDN, which themselves reiterated the findings of Stefan Robock and BNB staff during the 1950s.

SUDENE expected to finance less than 22 percent of its projects directly; the remaining funds were to come from other government agencies and private sources of aid and investment. Budgetary figures provided in the Guiding Plan indicate that industrialization was to be SUDENE's main concern initially: two-thirds of its predicted expenditures for 1961–65 were allocated for road construction and electric power generation, while only 14 percent was allocated for improvements to agriculture and the food supply. Another 16 percent was designated for public health and basic education, but few details were offered about such projects. SUDENE's public health endeavors consisted mainly of water sanitation, and the plan made vague references to "pioneering experiments in the field of basic education."[54] The agency lacked trained personnel in those areas, and its proposals encountered opposition from municipal governments that had been responsible for administering health and education programs (however inadequately).

SUDENE's primary focus on industry and infrastructure reduced political opposition to its activities. The agency hoped that industrialization would stimulate broader social and economic change "through the ampli-fication of the internal market, greater economic stability, larger income for the government, and, therefore, better public services. Only with industrial development will it be possible to modify the economic structure of the Northeast, facilitating the transition to a greater equality of income distri-bution and an economic system of greater internal dynamism."[55] Like Miguel Arrojado Lisboa's assumption in the 1910s that DNOCS's politically palatable focus on dams would inevitably lead to irrigation projects, Furtado and his staff saw industrialization as the first step in a larger program of regional reform. Their confidence in industry's ability to ignite widespread regional transformation proved to be unduly optimistic.

In trying to carry out SUDENE's first Guiding Plan, Furtado immediately confronted the scarcity of suitably trained technical staff in the northeast. As he recalled in a memoir, "We did a study of the social origins of the agronomists working in the region for the federal government, and we ver-ified that every last one of them came from families of large landowners and similar backgrounds. Certainly they were not the most suitable agents to conceive and execute an agrarian reform program."[56] These personal linkages to estate owners help to explain agronomists' reluctance in the 1930s through 1950s to significantly oppose elite interests. SUDENE began offering agronomy fellowships to students of modest background from the interior northeast and provided additional scholarships in technical fields like geology. The agency's goal was to retain good students in the north-east, since many who attended universities in the south-central region re-mained there to work, and to increase the pool of well-trained technical personnel from diverse social backgrounds.

## The Orós Dam Break

As if to confirm the need for new leadership in Northeast development, DNOCS faced an embarrassing disaster in March 1960. Kubitschek's flagship project in the sertão was Orós Dam; its construction had begun in 1922 under Epitácio Pessoa, but floods in 1924 caused considerable damage. A new loca-tion was selected in 1930, two kilometers from the original site, and plans to reconstruct the dam there lay dormant for almost thirty years. Kubitschek promised in his 1955 election campaign to make the long-awaited dam in Ceará's Jaguaribe River valley a reality. He asked DNOCS to resume work on Orós in October 1958, as a "work front" for drought refugees—with the goal of having it completed in less than two years. These were the same years in which Kubitschek's administration was overseeing the construction of a new national capital on Brazil's central plain—at tremendous expense. The Orós reservoir was intended to have a capacity of two billion cubic meters, almost half that of all other reservoirs administered by DNOCS at the time.

In early 1960 Orós was behind schedule, and DNOCS director Parente Pessoa pleaded with the MVOP to supply more money so the dam would be nearer completion when the rainy season began in March. Typically, funds budgeted by Brazil's legislature each year did not arrive until April. In 1960, disbursements were delayed until May, slowing Orós's construction. By late March, heavy rains threatened to overtop and erode Orós's unfinished packed-earth dam, which would set DNOCS's work back and send a dangerous volume of water into the populated valley below. In desperation DNOCS flew a helicopter over the valley to warn its 160,000 human inhabitants that they must move to higher ground. A few days later, flood waters poured over the dam, damaging eleven thousand properties below the reservoir.

Furtado and other representatives from SUDENE were present at the dam break, and the agency oversaw payments to all affected landowners. In the investigation that ensued, DNOCS and several *cearense* representatives to Brazil's legislature accused Kubitschek's finance minister of having redirected funds intended for Orós to two other ambitious projects: the Belém–Fortaleza Highway (connecting the northeast to the Amazon) and Brasília. They asserted that the dam had never been allocated sufficient funds to meet the accelerated timeline set by Kubitschek for its completion. The inaugural party for Brasília had a budget almost half that initially allocated to Orós before DNOCS realized that the project would require more money. (Ultimately Orós cost almost twice its initial budget, including expenses paid to compensate valley residents for flood damage.[57])

Furtado responded harshly to these accusations, saying that DNOCS was at fault for going over budget and falling behind schedule. Once it had become clear in early 1960 that a dam break was possible, the agency's engineers should have taken emergency measures to avert disaster, he contended. Although DNOCS's leaders and the MVOP testified that the 1960 flood was unprecedentedly severe, Furtado countered that similarly heavy rains occurred in the region every few years. Drought inspector Parente Pessoa complained bitterly to MVOP Amaral Peixoto that Furtado understood little about engineering or regional pluviometry and that Furtado's public criticisms of DNOCS were unfounded.[58] In his memoir about this period, Furtado recalls being shocked that the MVOP tried to cover for DNOCS's failings rather than initiating an investigation of the Orós fiasco. This was another important moment in Furtado's introduction to Brazilian political culture.[59]

Following the dam breach, DNOCS obtained additional funding and succeeded in completing the Orós project in nine months. At the dam's inauguration, Kubitschek's MVOP portrayed both the project's rapid completion and the rescue of people stranded in the valley after Orós broke as symbols of the government's renewed solidarity with Nordestinos.[60] A commemorative history of the dam, titled *Orós: Reservoir of Hope*, emphasized the efficiency and competence of the DNOCS engineers who oversaw its construction. It claimed that the earth-moving equipment required to complete Orós within two years used more power than that available from all the electric plants operating in the Amazon, northeastern, and central states at the time. Sixteen hundred men were employed on the project, and they and their families occupied a temporary town of eight thousand.[61] This heroic reading of Orós's construction appears to have been accepted by many ordinary northeasterners. Numerous *cordel* poems were written about floods and dam breaks, including that of Orós, but they typically refer to the inundations as acts of God, even punishments for human greed and ambition. Politicians and engineers are not depicted as being any more at fault than the rest of the population for such "natural" catastrophes.[62]

## Political Tumult in the Early 1960s

While the Orós drama was unfolding, SUDENE's Guiding Plan languished in Brazil's congress. Many northeastern representatives in the lower chamber opposed it, and the region's senators succeeded in blocking its approval. As Brazil's 1960 presidential election approached, Furtado scrambled to obtain the backing of presidential candidates from both ends of the political spectrum. He persuaded influential governors from the right-wing União Democrática Nacional (National Democratic Union) and left-wing Partido Trabalhista Brasileiro (Brazilian Labor Party) who supported SUDENE to obtain commitments from their parties' candidates that the new agency would have their backing. This effort, and Furtado's savvy decision to win northeastern governors' loyalty by including them on SUDENE's Deliberative Council, paid off. President Jânio Quadros, elected from the União Democrática Nacional, did not interfere with SUDENE. But to universal astonishment, Quadros resigned after seven months, due perhaps to frustration with the lack of authority given to the president under the 1946 constitution (though other speculations include mental illness). His independently elected vice president, João Goulart of the Partido Trabalhista Brasileiro—who had also been Kubitschek's vice president and Vargas's minister of labor—acceded to the presidency following Quadros's unexpected resignation. Goulart was an admirer of both SUDENE and Furtado.

In December 1961 the senate agreed to pass SUDENE's first Guiding Plan, but only with a series of amendments that significantly reduced the agency's potency as an instrument of reform. Workers in Recife sponsored a one-hour strike on December 6 in support of SUDENE. Shortly thereafter, Brazil's lower legislative house—in which the northeastern bloc had less influence (proportional to its share of the national population)—finally approved the Guiding Plan, without the senate's amendments.

In a report on SUDENE's accomplishments during its first few years, UN economic adviser Stefan Robock praised Furtado for his deft political maneuvering. By declining to align his agency with any one political party, Furtado obtained the support of competing politicians based on the soundness of his technical proposals. His political base included the governors of northeastern states who served on SUDENE's Deliberative Council; a growing reformist coalition in the northeast, including liberal bishops; southern politicians who saw SUDENE as a way to end corruption and inefficiency in aid provision to the northeast; Brazilian intellectuals and university students who embraced Furtado's nationalist development ideology; and the U.S. government, which sought forces for progressive change in Brazilian society that might diminish the appeal of the supposedly communist Peasant Leagues. Furtado's political acumen distinguished him from prior *técnicos* like DNOCS agronomist Duque, who had tried to alter regional development priorities for many years with only modest success. Whereas Duque's admirers described him as a conscientious *filósofo*, a man devoted to his analyses of crops and soils, Furtado possessed a rare combination of technical prowess and maturing political insight that made him a much more influential figure. Robock described Furtado's first plan for northeastern development, submitted to Kubitschek in 1958, as "a political document cloaked in the authority of technical economic analysis."[63]

Despite his admiration for Furtado, Robock voiced several criticisms of the Brazilian economist's work. First, he felt that SUDENE tried to produce dramatic results too quickly in order to justify its preeminence over DNOCS and other agencies that questioned the need for another organization dedicated to northeastern development. As a result, Robock thought, SUDENE focused on improvements to physical infrastructure at the expense of investment in human resources, because there were few studies available suggesting how to improve regional education and public health. In keeping with the approach to modernization embraced by the U.S. Department of State during the administration of President John F. Kennedy (particularly under the Alliance for Progress), Robock believed by the early 1960s that attention to human resources was critical for northeastern development. Since mechanized farming would make many farm workers redundant, the most important step was "to make the long-run solution of moving people out of agriculture well understood, widely accepted, and reasonably attractive" through public outreach and improved education.[64] The remaining farmers would require extensive training in modern cultivation methods, Robock advised. He also criticized SUDENE for executing its Guiding Plan in a top-down manner, despite its purported support for cooperation across agencies. This was an unfortunate consequence of the superintendency's rush to produce and pursue a new regional strategy, but it further alienated DNOCS in particular.

Robock disagreed with a core premise of Furtado's agenda—namely, that Brazil's disparate regions needed to be brought into greater equality. He saw regional disparities as inevitable, and recommended working toward overall national growth and accepting regional inequalities as a consequence of rapid development in some sectors.[65] This view was anathema to regionalists like Furtado, for whom development was not simply about encouraging national economic growth but was fundamentally about redressing injustices that had persisted in Brazil, and particularly in the northeast, since the colonial era. As a result, Furtado placed greatest emphasis on industrializing the northeast so it would catch up, in wealth and consequent political importance, with southern Brazil.

In 1963 President Goulart made Furtado his minister of planning and asked the economist to draft a three-year development plan for the entire country. Inflation, which reached an annual rate of 52 percent in 1962, had eroded the economic optimism of the Kubitschek era. Brazil was suffering from reduced foreign investment, the result of IMF concerns about government overspending and investors' nervousness about left-wing radicalism in Latin American (triggered particularly by Fidel Castro's alliance with the Soviet Union). Furtado laid out a strategy for curbing inflation while continuing to invest in development by raising taxes and reducing subsidies to successful industries. His agenda did not sacrifice social reforms to the need for economic growth, since he believed development had to benefit all social sectors for Brazil to prosper. As a result, his plan demanded sacrifices from the wealthy and faced opposition from many quarters. In May he returned to SUDENE, recognizing that his reforms would be rejected by Goulart's opponents in the legislature.

In mid-1963 SUDENE obtained approval for a second Guiding Plan containing more specific proposals than the first. Furtado's memoir portrays the subsequent months as a period of tremendous energy and accomplishment as SUDENE's staff began to reap the fruits of their initial studies and undertakings. Supported by Goulart, the agency continued to promote agricultural reform. A report by the Ministry of Agriculture concluded that redistributing large estates to small farmers would increase farm productivity, provided that the farmers had access to credit and the basic medical and sanitation infrastructure necessary for their own well-being.[66] SUDENE assisted small farmers in forming cooperative associations to qualify for bank loans. In June 1963 a law was passed (no. 4, 229) permitting DNOCS to establish "rural centers" of small farms along river banks if that land was not being efficiently farmed by its existing owners—similar to what CODENO had advocated in 1959.[67] DNOCS would share the cost of constructing irrigation canals with the settlers and provide other support such as electricity, health services, basic education, seeds for planting, and crop storage facilities.

With the aid of their growing unions, rural workers won legal rights in 1963 similar to those that had been granted to urban workers under Vargas in the 1930s. These established a variety of labor protections, permitted participation in unions regulated by the government, and provided legal mechanisms for redressing grievances. In the context of these dramatic victories for the political left, Goulart's support for land redistribution, along with his proposals to reform Brazil's tax system and extend voting rights to enlisted soldiers and illiterates (who made up nearly a quarter of the northeast's population), made moderate and conservative elements in the Brazilian polity increasingly uneasy. The ideological cleavages between Brazil's major political parties intensified, and Goulart found himself under attack from both right-wing enemies of his reformist agenda and left-wing idealists disillusioned with what they perceived as pandering half measures.

One powerful organization wary of Goulart was the U.S. Department of State. News of the Peasant Leagues had reached the U.S. press in 1960, adding fuel to Cold War fears of communist uprisings throughout Latin America. State Department staff in Brazil were limited in their perception of Brazilian politics because most did not speak Portuguese. They relied heavily on English-speaking contacts who represented the most highly educated and wealthiest segments of Brazilian society, particularly in the northeast. Based on the information made available to them through these sources, State Department representatives believed that Brazil was vulnerable to a communist takeover.[68] To counter this, the United States established an Alliance for Progress mission in northeastern Brazil, in hopes of generating popular support for politicians more palatable to the U.S. government than radical leftists.

The Alliance for Progress initiative, which had been created by President Kennedy following Fidel Castro's rise to power in Cuba, focused on highly visible projects that made the recipient population aware of American aid efforts. These included water purification, rural electrification, mobile health units, and construction of elementary schools. Northeastern Brazil was one of the first targets of Alliance for Progress assistance. Furtado met with Kennedy in July 1961 to discuss plans for the region, and he received a commitment of an American advisory mission that would evaluate the northeast and propose an aid package. In April 1962 SUDENE and the U.S. Agency for International Development formulated an agreement concerning their respective goals. By June of the following year, 133 Alliance for Progress staff were working in Recife.

Relations between SUDENE and the Alliance for Progress soon soured. SUDENE was wary of American intrusion onto its terrain and disapproved of the alliance's focus on short-term projects that could quickly influence popular opinion. U.S. State Department personnel stationed in the northeast increasingly saw their primary goal as combatting communism. They angered Furtado by providing aid directly to conservative northeastern state governors and gubernatorial candidates whose politics the United States approved of. Furtado claimed that American funds should be funneled through his organization, since Brazilian law required foreign aid to be negotiated at the federal level. But Goulart chose not to oppose the American effort, even though it was clearly aimed at weakening his administration. Alliance for Progress personnel seemed blind to the turf battle that their presence had provoked, and they interpreted SUDENE's uncooperativeness as ideologically motivated obstruction.[69]

An American observer of the Alliance for Progress mission in Brazil during the early 1960s believed its staff had profoundly misinterpreted SUDENE's motivations. Looking back from the vantage point of the 1970s, Riordan Roett concluded, "In failing to work with the SUDENE, the United

States . . . lost an opportunity to strengthen not only the social and economic fabric of the [Northeast] but to support the important political aspect of societal change," which Furtado understood.[70] Left-wing governor
Miguel Arraes of Pernambuco, whose political ambitions the Alliance for
Progress hoped to thwart, described the American government's misperceptions about northeastern Brazil in a *Newsweek* interview published in
March of 1963: "You are only giving us chocolates and candy, while what
we need is jobs. You talk of us as if we were an international menace, and
what we are is a poor region full of suffering and human problems. What
we want is very little—your understanding. But you behave like those soldiers in *The Teahouse of the August Moon*—you insist on making us into
something we cannot be."[71]

By 1964, support for Goulart's agenda was eroding throughout Brazil.
In the face of a stagnant economy and spiraling inflation, the president
found himself under siege from opponents of all political stripes. In an attempt to retain the support of workers, enlisted men, and university students, he staged a rally in mid-March during which he decreed the federal
expropriation of "underutilized" property in several categories. This was
widely seen as a decision to align himself with the radical left, including
the reforms demanded by Francisco Julião and the Peasant Leagues.[72] For
many in the middle class who feared a revolutionary uprising similar to
what was occurring in Cuba this was intolerable. They staged demonstrations asking the military to exercise its traditional "moderating" power by
overthrowing Goulart's administration. On March 31, 1964, Brazil's military launched a coup that brought down the elected government with little
resistance. The military dictatorship would remain in power for more than
twenty years.

After removing or intimidating leftist officeholders, including Goulart, the
generals in charge of Brazil's new government had successors elected by the
remaining state and national legislators, all of whose political views were acceptable to them. The new regime closed down left-wing presses and confis-
cated books that they deemed incendiary from stores and private homes.
They made the Peasant Leagues illegal and arrested many of the movement's
leaders. Rural landowners read this as a sign that they could take revenge on
rebellious workers, though military leaders soon stipulated that peasants'
rights—including the right to organize in state-sanctioned syndicates—
should still be recognized.[73] The generals abolished "consciousness raising"
literacy programs in the northeast led by progressive Catholic priests and
Marxist pedagogue Paulo Freire, because they perceived such programs to be

politicizing the poor. While purging Brazil of such supposed threats to social stability, the new government publicized any evidence it could find of a narrowly averted communist uprising. Tangible signs of a pending rebellion amounted to nothing. The most damning evidence was a collection of ten thousand overalls found in a warehouse owned by a Pernambucan land reform agency, which the army and police claimed were the intended uniforms of a peasant militia, but were probably meant for distribution to smallholders who settled the organization's irrigated colonies.[74] Nonetheless, U.S. ambassador Lincoln Gordon formally commended Brazil's generals for their bravery in saving the country from communist insurrection.

When the coup began, Celso Furtado closed SUDENE until further notice. Humberto de Alencar Castelo Branco, the new military president, had the power to suspend the political rights of state enemies for ten years, and Furtado was one of many prominent figures subject to this censure. (Pernambucan doctor Josué de Castro was another.) As is evident in Pernambuco's secret police files, Furtado's success as an advocate for change led the military to view SUDENE's staff as subversives.[75] Furtado went into exile in Chile, the United States, and finally Paris, where he remained a professor at the Sorbonne for over twenty years. The renowned technocrat's hopes for transforming his native region were shattered. Furtado's subjection to censure confounded even his American critics, who recognized him as one of Brazil's most impressive intellectuals and a highly capable bureaucrat. Nontheless, when SUDENE reopened under leadership approved by the military, the United States increased its aid to the agency. Unsurprisingly, SUDENE dropped the more radical elements of it agenda and focused primarily on industrialization, infrastructural improvements, and agricultural modernization in the northeast. Its greatest successes were in the areas of electric power supply and water sanitation.

In the years that followed, Brazil's military government lost interest in the northeast. During the 1970s the dictators turned their attention to populating and developing the Amazon; construction of the Transamazonian Highway was among their most significant projects in this regard. The government hoped to avoid confronting the contentious issue of land ownership in the northeast by opening the Amazon to industry. Underemployed workers from the sertão were encouraged to seek employment farther west, in Brazil's last great frontier. SUDENE's funding decreased, and the U.S. Agency for International Development's program in Brazil adopted a national rather than regional focus, in keeping with the military government's more centralized administration.

## Progress Undone

In the face of two droughts during the 1950s that discredited DNOCS as a development agency, economists asserted themselves as essential participants in regional planning. To bolster their claims to development expertise, this new technocratic cohort cited the central role played by economists in establishing the TVA's integrated planning model. Yet more forcefully than their American advisers, Brazilian economists working in the northeast proposed reforms to the social organization of production as well as to technical infrastructure. Their leader, Celso Furtado, argued that the region's coast and hinterland must be examined as an integrated economic system and that longstanding labor and landholding patterns had contributed to northeastern underdevelopment. Whereas most engineers and agronomists acknowledged the social dimension of sertão "backwardness" only once they had spent months in the interior watching poor families struggle with disease and starvation, Furtado entered the national debate well attuned to the social and political dimensions of the region's challenges. He proved unusually deft at maneuvering among political factions; he placated elites and military officers for a time in the early 1960s while adopting a limited agrarian reform agenda that promised genuine improvements to the security of sertanejo farm workers and the regional economy.

In keeping with the development theory espoused by his colleagues at ECLA, Furtado assumed that a number of progressive social changes would follow from economic growth in the northeast. He saw industrialization as the first and most politically feasible step in a broader reorganization of the regional economy, one that would provide a greater range of employment opportunities for the growing population. Conservative sectors of northeastern society embraced new industries and a degree of agricultural mechanization, but these elements did not provide greater wealth or political independence for the sertanejo poor. Absent the essential components of land reform and irrigated smallholding, no development measures could make tenant farm workers or smallholders resilient to drought. The near passage of these more radical measures under Goulart helped to promote a conservative backlash. In the aftermath of the military coup, economists like Furtado were as impotent as the series of technocrats who had preceded them in the drought zone; they could not force legislation that elites concertedly opposed. In some ways Furtado was less moderate in his political ambitions than his predecessors in sertão development, and he was ultimately the most dramatically thwarted. Several years after Brazil emerged from military rule, Furtado published a memoir of his experience with the "northeast operation" to encourage a new generation of Brazilians who wished to redeem their country, and to give them the benefit of his own political education during that time.[76]

# Conclusion

## Science, Politics, and Social Reform

........................................................

### The Generals' *Sertão* Legacy

Following the 1964 coup, José Guimarães Duque continued to speak of agronomists' work in social and moral terms. Duque advised his staff to exercise humility, recognizing that the people they assisted were "creations of God, part of a family, citizens," with strengths and deficiencies like all human beings.[1] But the quarterly bulletin of the Departamento Nacional de Obras Contra as Secas (DNOCS; National Department for Works to Combat Droughts) in which views of this kind were promulgated, ceased to be published in 1966; it was replaced by the *Boletim Técnico* containing almost exclusively quantitative data pertinent to civil engineering, hydrology, and related geological sciences. Two years later, DNOCS and the Superintendência de Desenvolvimento do Nordeste (SUDENE; Superintendency for Northeast Development) were placed under the authority of a new Ministry of the Interior. Over the following decade, the military government established several projects aimed at developing the most dynamic sectors of the northeastern economy and providing credit for investment in farms and ranchland. These included the POLONORDESTE development program (1974); Projeto Sertanejo, focused specifically on the semiarid sertão (1976); the Companhia de Desenvolvimento do Vale do São Francisco (Development Company of the São Francisco Valley), for irrigation; and the PROTERRA credit program. In a detailed analysis of these efforts published shortly after the dictatorship ended, Otamar de Carvalho described the regime's neoliberal approach to development as "conservative modernization."[2] In the absence of attention to class inequalities, and with no serious effort to reorganize land ownership, the bank credit and technical infrastructure offered as part of the above programs merely solidified the advantages of the landowning class over their workers. Developers focused more on export crop marketability and regional economic growth than on the social impact of their projects. The years of military rule recapitulated, and perhaps intensified, the pattern that had developed over the drought agency's first half-century.

Agronomists working for DNOCS and other federal agencies, such as the Empresa Brasileira de Assistência Técnica e Extensão Rural (EMBRATER, Brazilian Enterprise of Technical Assistance and Rural Extension, an agricultural extension service established in 1975) still hoped that their irrigated settlements would serve as centers for economic and social development, with health posts, schools, and roads aiding many more people than the *irrigantes* who were selected as direct beneficiaries. Yet construction of irrigation works proceeded slowly. By the end of 1977 just over two thousand families occupied smallholder colonies administered by DNOCS, the majority of which were in the state of Ceará. The fourteen thousand irrigated hectares in these settlements (representing just over half of the colonizers' farmland, the remainder being "dry areas" for ranching and xerophilous crops) amounted to about 12 percent of the agency's planned irrigation network. The other 88 percent was yet to be constructed.[3]

Scholars who analyzed DNOCS's irrigated colonies in this period note that many residents (*moradores*) from whom land had been expropriated resisted participation in the new settlements because they resented the federal government's claims on their land, homes, and other property. For those who declined to become irrigantes, land speculation and rent increases arising from planned infrastructural improvements made it difficult to relocate nearby, given the modest compensation provided for their losses. Men who passed the drought agency's selection process to become irrigantes described themselves as "subjugated" by colony managers' requirements. Many expressed nostalgia for their prior independence.[4] The selection process for irrigantes involved interviews with other area residents about a candidate's reliability, community-mindedness, diligence and productivity, family harmony, religion (Catholic, Protestant, or "other"), and vices. Results were allotted points, with the highest ranking going to men between the ages of twenty-two and thirty-five whose household comprised at least ten members, including additional males between the ages of fifteen and forty-five.[5] Presumably some former residents of the colonized areas were not selected, even if they were willing to participate.

In the mid-1970s scholar and policy analyst Anthony Hall studied three irrigation projects on which four hundred families were settled (representing roughly 40 percent of all households in a drought agency–administered colony at that time). On the whole Hall found that the costs of establishing these settlements outweighed their social benefits. The colonies displaced families from desirable land in relatively fertile valleys, and the selection criteria worked against the farmers in greatest need of assistance. According to one state university economist, the average income of families farming irrigated settlements was less than 75 percent of the minimum wage in Recife.[6] Hall proposed other approaches to sertão development that he believed would be more cost-effective and socially beneficial. These included collective rather than individual farming (something the right-wing government was wary of, fearing a resurgence of the Peasant Leagues); providing better technologies, bank credit, and market access for existing farms rather than displacing families to establish new settlements; and incentivizing relocation to underpopulated areas in or near the sertão that could be successfully farmed (with appropriate government assistance) without recourse to expensive irrigation networks. The latter proposal, he noted, pitted humble sertanejos against influential ranchers who could access PROTERRA credit to expand into these areas, and thus were a political obstacle. On the whole, Hall viewed irrigated colonies as pet projects of technocrats and some landowners but not the best use of resources to bring food security to the most marginal sertanejos.[7]

More than a decade after the 1964 coup, one of Celso Furtado's former assistants, Francisco de Oliveira, published a critique of SUDENE's accomplishments under military rule in which he accused the organization of having functioned primarily to stabilize the northeast's agro-ranching economy.[8] Following the coup, SUDENE's appointed directors were members of landholding families intent on promoting their own class interests, which had grown to encompass banking, transportation, and manufacturing as well as agriculture. Even projects that relocated sertanejo small farmers to new settlements on the Amazon frontier served the needs of the rural elite, Oliveira insisted. Estate owners became more willing to invest in intensive farming as DNOCS increased its provision of irrigation canals and mechanized technologies. State-funded resettlement provided employment for their former workers, whom DNOCS also expected to adopt new farming methods, though with little training. Extreme inequalities in landholding persisted: in the mid-1970s, when the sertão population was roughly twelve million, 58 percent of northeastern land area accounted for a mere 8 percent of property holdings. Soil degradation on the smallholdings occupied by the majority of sertanejos intensified those families' poverty, and young adults continued to migrate to cities in search of a more secure future.[9] SUDENE remained focused on the northeast's most dynamic economic sectors, aiming to diversify the regional economy.

· · · · · ·

The thrust of most criticism leveled at DNOCS and SUDENE once Brazil's dictatorship ended in 1985 is that the military's development priorities for the northeast favored industry at the expense of rural workers. Small farmers were displaced to build dams whose waters were used primarily for hydroelectric power that was directed to cities. The result was a continuing scarcity of food crops and economic opportunity for the rural poor. Placing primary emphasis on their needs would have dictated a very different menu of development projects—for example, providing cisterns for household water storage and securing land rights.[10] Nutritional studies conducted during the dictatorship found evidence of Nordestinos' inadequate nourishment in their weight and height measurements, which were low relative to those of southern Brazilians. Dietary analysts noted a substantial drop in caloric intake among workers in Recife from 1960 to 1975, due to the increasing scarcity and resulting high price of food.[11]

Throughout his career, esteemed Pernambucan geographer Manuel Correia de Andrade advocated adopting an agricultural regime in northeastern Brazil suited to its social and climatic realities. In 1970 he wrote, "The nonexistence of an agrarian [land reform] and irrigation policy for this area has contributed to its subutilization and, consequently, to the loss of resources that might have mitigated the circumstances of the regional population during periods of drought. The employment of high-tech [farming] methods, without regard for the social order, has led to the subutilization of large areas."[12] Small farmers had limited ability to adopt modern agricultural technologies without the option of pooling their resources in cooperative organizations. A decade later, Andrade still felt that planners had neglected to view the Northeast's rural and urban economies as an interconnected system that should be attended to simultaneously—with an eye to broad social improvements. He believed that development policies for the Northeast had been most favorable to industrialists, many of whom were not themselves Nordestinos. Ranchers were the second most benefited group, with urban and rural workers falling well behind.[13]

Reflecting on SUDENE's legacy at a conference in 1984, Furtado noted that although the northeast had obtained significant infrastructural improvements over the two decades of military government, the region's overall economic growth was not reflected in the inflation-adjusted incomes of most Nordestinos. Without attention to subsistence farming and the needs of small farmers, he warned, "policies to improve agriculture tend to degenerate rapidly into policies to create surpluses that benefit privileged groups."[14] In order to raise workers' average income, Furtado recommended that the government limit its subsidies to industries that provided expanded employment opportunities in the northeast. In a later critique, the economist described the region's agricultural and industrial sectors as locked in a vicious cycle of underdevelopment. The dearth of food crops forced the importation of food from the south. This was expensive and required industries to pay higher wages, thus retarding industrialization. In Furtado's opinion, only the existence of an "escape valve" for desperate Nordestinos, in the form of southern Brazil's largest cities, had prevented a mass uprising against self-serving elites.[15] This migration pattern aided southern industrialization by increasing the labor pool there, but at a significant cost to urban residents in congestion and crime—most acutely in São Paulo.

More recent observers of conditions in the sertão argue that access to drought aid still relies on patronage relationships, leaving poor sertanejos in a state of "persistent vulnerability" to the vagaries of climatic and political fluctuations. The multiyear drought of 1997–1999 was reported to have left ten million people (roughly half of the sertão's population) "on the brink of starvation" after the first year.[16] Families tried to survive on remittances sent by relatives who had migrated to São Paulo and on erratic government aid—generally food rations or cash wages paid in return for labor on public works projects (a practice begun in the 1920s). Land that remained arable was converted to forage (to sustain ranchers' cattle) rather than food crops. These descriptions of drought-induced crisis and response at the turn of the twenty-first century are eerily similar to accounts from eighty years earlier. Anthropologists Donald Nelson and Timothy Finan argue that sertão development should focus on residents' capacity to resist drought independently, diminishing their reliance on state-sponsored emergency assistance.[17] Many sertanejos endure what one might call drought citizenship, in which their relationship to the state is mediated primarily by their vulnerability to climate crisis.

There are some hopeful studies of late twentieth-century trends in sertão development, but these appear to be glimmers of hope rather than sustained improvement. Social scientist Judith Tendler observed a promising democratization of drought aid in 1987 led by Ceará's agricultural extension service, in which agronomists backed by the state bureaucracy organized community councils to determine which projects would be most helpful to those in greatest need of aid. This process helped to reduce clientelism in the distribution of government assistance, and it established clear expectations that the needs of the poorest residents would be prioritized. But the approach was short-lived, and when another drought occurred in the early 1990s, local mayors and landowners had regained control over disbursements.[18] It seems likely that the brief progress made in 1987 reflects the democratic fervor witnessed throughout Brazil immediately after the military dictatorship ended, since the inertial force of long-standing power structures was reasserted soon after. In another evaluation of changes to development ideology following democratization, Robert Silva notes the emphasis on "living with" the drought rather than "combatting" it. He sees this rhetorical shift as indicative of a more ecologically minded view of interactions between humans and nature, different from the modernist assumption that landscapes should be reengineered to meet human needs.[19] But in recent decades, the overuse of fragile soils by sertão smallholders has made the impact of even climatically minor droughts worse.[20] Concerns about sustainable farming and soil conservation have led to some lower-tech projects to meet the needs of farmers scattered throughout the sertão, such as constructing household cisterns.

Future historians can evaluate the significance of the emphasis on *convivência* (coexistence) in shaping the work of twenty-first-century sertão development agencies. On the whole, there is no indication of a dramatic shift in sertão development strategy from the dictatorial period onward. DNOCS continued to focus heavily on dam construction along with reservoir management and maintenance (see table). In 2002 the agency completed the Castanhão Dam in Ceará, creating a reservoir with a capacity almost quadruple that of the Orós Dam (which was the largest reservoir during the agency's first half century by a considerable margin). A centennial publication by DNOCS's Ceará division (the agency's most significant) reported that by 2009 slightly over four thousand families were settled in fourteen irrigated colonies within the state, with plots averaging about five hectares per household.[21] This is well below José Guimarães Duque's ambitions in the 1950s (though, based on the critiques by Anthony Hall and others, that may be for the best). Other recent projects include large pipelines to transport water beyond the land immediately adjacent to reservoirs; discussions of an ambitious "transposition" of the São Francisco River to distant areas of the sertão are ongoing. Under the administration of President Luiz Inácio Lula da Silva (2003–11) the principal strategy for poverty

Capacity of Reservoirs Completed in Ceará

*1906–1962*

4,805,941,000 m³ in 34 reservoirs administered by DNOCS (including
1,940,000,000 m³ in Oròs)

57,120,000 m³ in reservoirs built by the agency but administered by another
organization (e.g., municipalities—mostly constructed 1910–22)

735,551,000 m³ in 335 reservoirs on private property (averaging 2,195,674 m³
each)

*1965–2006*

10,021,673,000 m³ in 30 reservoirs administered by DNOCS (including
6,700,000,000 m³ in Castanhão and 1,601,000,000 m³ in Arrojado Lisboa)

3,860,000 m³ in a reservoir built by DNOCS but administered by the
municipality of Boa Viagem

221,990,000 m³ in 131 reservoirs on private property (averaging 1,694,580 m³
each)

*Note*: m³ = cubic meters.
*Source*: Data compiled from DNOCS publications.

alleviation was the Bolsa Família (Family Purse). This provided "conditional cash transfers" to poor families that could be used to secure access to food and incentivized sending children to school, among other things. These efforts seem to have absorbed most of the "Lula" administration's attention to marginal populations in the former president's native region.

## Blind Spots in the Technocratic Lens

Throughout the twentieth century in northeastern Brazil, technocrats employed by development agencies expanded hydrologic and transportation infrastructure in ways that improved the security of export agriculturists and ranchers but did little to increase the economic and food security of those who suffered most during droughts. Many elements of this history are relevant to contemporary debates about adaptation to the realities of climate change. As Amartya Sen and others have argued, vulnerability to so-called natural disasters is highest among already marginal populations; overemphasizing the "natural" causes of their suffering deflects attention from the social dynamics that perpetuate the insecurity of the poor.

The history of drought aid in northeastern Brazil can be read as a parable of technocratic development more generally. For much of the twentieth century, men trained in a range of technical fields offered solutions to entrenched poverty that attracted politicians wary of disturbing the prevailing social order. Organizations like the Rockefeller Foundation's International Health Board, which pioneered applying scientific knowledge to achieve social transformation, believed that their staff had rational solutions to pressing problems of disease, hunger, and poverty. In the view of many politicians and aid agencies, the advantage of pursuing social change through science was that science and its practitioners were understood to be apolitical; technical knowledge in fields like medicine, engineering, and agronomy applied similarly in all social contexts. Given the high stakes involved in mitigating poverty in the postcolonial world, with revolution, mass redistribution of property and a profound realignment of political power as one possible route to change, governments embraced science because it offered a more manageable path to modernization and social betterment.

Yet scientists, and technicians using applied scientific knowledge, operate in social and political landscapes that shape the impact and effectiveness of their work. In the Brazilian sertão, four cohorts of technocratic actors found themselves with limited ability to address the humanitarian crisis caused by drought over several distinct political periods. Public health crusader Belisário Penna set the stage for the experts who followed him by establishing an overtly political discourse about sertanejo marginality in Brazil's First Republic. His speeches and essays are an extended harangue against the self-indulgent leadership of an oligarchic class that remained indifferent to the misery of its poor rural compatriots. Penna envisioned a utopian future in which Brazil would be guided by scientific thinkers like himself—sage men of substantial education whose policies would promote all citizens' welfare in the name of national progress. This future was not realizable in a country where political power stemmed primarily from wealth grounded in landholding.

During the period in which Penna was writing, and for several decades afterward, three other coteries of technocratic developers used their expertise to aid the rural population that Penna identified as most abandoned by the state: sertanejos in the semiarid northeast. The first of these groups, civil engineers working for the federal drought agency during the First Republic, were limited in their ability to effect social reform partly by their own training and professional allegiances. They needed the support of regional elites to sustain their agency, and Nordestino elites wanted infrastructural improvements without alteration to the social order. Telegrams and other archival evidence reveal that engineers working at sertão construction sites were often persistent advocates for their labor force, palpably aghast at the human misery confronting them. Yet their primary obligation was to oversee construction, and they focused their efforts nearly unwaveringly on this—however unsavory it may have been to build reservoirs and roads on the backs of the starving.

The drought agency's managing engineers convinced themselves that the dams they constructed would be the first step in an evolving process of modernizing sertão agriculture and food production. They clung to assumptions about the social utility of reclamation efforts elsewhere in the world, particularly in the western United States, as verification that dams could form the cornerstone of a democratic agrarian society in semiarid terrain. In retrospect, one could argue both with the engineers' understanding of reclamation's impact in other semiarid places and with their identification of climatic similarities as core features linking the sertão to regions with very different political and social histories. As economists would later argue, focusing on the drought per se as the sertão's fundamental problem distracted from more basic issues of political and social inequity in the northeast. Yet the focus on climate served civil engineers' interest in promoting their own expertise as agents of modernization, and it allowed the elites who supported them to obtain what they most wanted from the federal government: funding for reservoirs and roads.

Agronomists hired by the drought agency from the 1930s onward were limited by their minimal professional authority and marginal bureaucratic position within the drought agency. Their profession received government recognition in Brazil only during the first administration of President Getúlio Vargas, and practitioners consequently had no history as state development advisers when the drought agency's agricultural service was established. Agronomists generally remained below engineers in DNOCS's management hierarchy, and engineers had their own reasons for promoting dam and road building over irrigation networks and agricultural extension programs. Additionally, agronomists often worked in a political climate hostile to politico-economic analyses of food security. During Vargas's repressive Estado Novo, in particular, government bureaucrats were encouraged to recommend purely technical solutions to social problems in keeping with Vargas's moderate, paternalistic approach to reform. Leading agronomists from this period framed sertanejos' vulnerability to drought as a problem of insufficient education, which fit comfortably with Vargas's largely rhetorical emphasis on incorporating northeastern backlanders into national modernization.

The men who administered DNOCS's agricultural extension services saw the benefits to be gained from expanded smallholding in the sertão, but they were in no position to aggressively pursue such a politically contentious measure. Instead they focused on acculturating sertanejos to irrigated farming and other technologies for more intensive cultivation, and they made little distinction between the classes of sertanejo farmers who sought their help. Ultimately, with very limited funding for their operations, the sertão's twentieth-century agronomists had little impact on the drought zone's social landscape. Commitment of federal aid to the region was fickle, and engineers kept most of what arrived for their own projects.

The last cohort considered in this study, development economists, identified closely with a growing international coterie of regional planners. Influenced by advisers from the Tennessee Valley Authority and by models promulgated by the UN's Economic Commission for Latin America, Brazilian economists during the 1950s felt empowered to assert that the entire thrust of drought aid had been misguided since DNOCS or its predecessor, the Inspetaria de Obras Contra as Secas (Inspectorate for Works to Combat Droughts), was first established. The sertão's problem, they contended, was not drought but poverty and inequality, and both were linked to regional underdevelopment. The solution was to reorganize the entire northeastern economy with an emphasis on industrialization and food security for workers. As with the strategies proposed by engineers and agronomists, this stance was professionally expedient. Economists aimed to displace engineers as the leading agents of regional modernization. And what better way to do so than by proposing a new orientation for development that they themselves were best suited to carry out? But their emphasis on underdevelopment also stemmed from an ideological conviction, forcefully articulated by Celso Furtado, that numerous political and social imbalances had to be redressed for the northeast to prosper.

Economists benefited in the 1950s from presidential administrations committed to ambitious modernization agendas, and sympathetic to Furtado's emphasis on industrialization as the starting point for regional development. Yet they soon confronted the limits of reformist technocracy's capacity to override conservative political opposition. Events during the mid-1960s revealed that fear of social upheaval was not confined to Brazil's rural elite but was shared by a variety of middle sectors, including conservative Catholics and small-scale property owners. At a volatile time in which profound reorganization of northeastern landholding seemed momentarily possible (certainly in the imaginations of a wary middle class), economists' relatively moderate recommendations for improving food security were too readily associated with more radical redistributive agendas, such as those of the northeast's Peasant Leagues.

· · · · · ·

Much historiography of twentieth-century science and medicine in the United States, and in regions under American or European imperial control, has portrayed science as a powerful apparatus for the exercise of state power. This view of science-as-behemoth has undergone some revision even in those national and regional contexts.[22] In the case of Latin America, where the history of science is a less established discipline, it is difficult to see scientists as having been empowered to effect significant social change— except in the case of particularly aggressive public health campaigns.[23] In Brazil's sertão, the efforts of the federal drought agency initially seem to exemplify massive state restructuring of a landscape in the name of social progress. Annual reports, produced sporadically throughout the drought agency's history, sketch numerous dams under construction and describe far-reaching plans for agricultural and public health improvement across the region. But what appears on paper to be an ambitious development program in the high modernist tradition characterized (and maligned) by James Scott turns out to have been largely an illusion of progressive action.

Many bureaucracies were created in Brazil from 1909 to 1959—and beyond—to address the misery caused by drought. A large number of earthen and concrete reservoirs of differing sizes were built, particularly in Ceará, to retain water in the sertão. But the impact of these efforts on sertanejo society was to solidify existing social relations, reinforcing landowners' control over natural resources and the human beings who depended on them and thus increasing landowners' power as local patrons. Contrary to many narratives of this kind, the actions of the federal drought agency and its successors did not significantly increase state authority in the northeastern hinterland. Federal agents working for development organizations remained largely beholden to traditional rural power brokers, and the priorities of the elite *coronéis* heavily influenced their agendas. When in the mid-1960s technocratic reformers backed by the president and some like-minded intellectuals had the gall to propose legislation that significantly threatened landowners' interests, Nordestino conservatives helped persuade other sectors of the national polity to overthrow both the president and the development apparatus that his administration supported. Technocratic expertise clearly did not trump the material authority of landowners and industrialists. Claims based in scientific analysis were put to various political uses when reigning power brokers saw their utility, but they had little capacity to influence political action simply on the merits of their grounding in empiricism and relevant professional experience.

Achievements in many scientific fields during the nineteenth and twentieth centuries contributed to euphoric hopes that a multitude of social ills could be cured by applying this new knowledge, and the technologies that relied on it, to intractable problems. Public health workers, engineers, agronomists, and others dedicated themselves to addressing the needs of populations around the world that had not benefited from numerous social advancements. These efforts were well intentioned, and in political contexts amenable to significant social change they could help to achieve it, but the social context was critical to the success or failure of these endeavors.

In the Brazilian sertão, as in other regions of entrenched poverty and inequality, many men with scientific training sincerely believed that their expertise was sufficient to rescue humble farm laborers from their travails in an unforgiving landscape. As a number of these technocrats came to realize, however, there was no way to accomplish this without engaging in high-stakes political conflict over scarce and essential resources: farmland, food, and water. This would pit them against intransigent and powerful regional elites, and often against the interests of their own profession and social class. Without sufficient political will to effect dramatic regional transformation, technocratic expertise had limited impact on sertanejo society. Drought remained a crisis not because of a dearth of technical knowledge but because of the losses that many influential people might have experienced if impoverished sertanejos' vulnerability were adequately addressed. It is this dynamic that makes Brazil's drought saga a parable of twentieth-century technocratic development.

# Notes

## Abbreviations Used in the Notes

| | |
|---|---|
| ABP COC | Arquivo Belisário Penna, Casa de Oswaldo Cruz, Rio de Janeiro |
| AN | Arquivo Nacional, Rio de Janeiro |
| APP | Acervo Público de Pernambuco, Recife |
| CPDOC | Centro de Pesquisa e Documentação de História Contemporânea do Brasil |
| FCRB | Fundação Casa de Rui Barbosa, Rio de Janeiro |
| FGE | Fundo Governador do Estado |
| FGV | Fundação Getúlio Vargas, Rio de Janeiro |
| IHGB | Instituto Histórico-Geográfico Brasileiro, Rio de Janeiro |
| RAC | Rockefeller Archive Center, Tarrytown, NY |

## Introduction

1. Ervin, "1930 Agrarian Census."

2. Sen, *Poverty and Famines*; Ferguson, *Anti-Politics Machine*; Escobar, *Encountering Development*; Scott, *Seeing Like a State*; Mitchell, *Rule of Experts*.

3. Cueto, ed., *Missionaries of Science*; Stepan, *Picturing Tropical Nature*; McCook, *States of Nature*; Rodriguez, *Civilizing Argentina*; Medina, *Cybernetic Revolutionaries*.

4. Carey, *In the Shadow of Melting Glaciers*; Wolfe, *Watering the Revolution*.

5. Rogers, *Deepest Wounds*.

6. Li, *Will to Improve*.

7. Davis refers to "triple peripheralization" of the northeast region. Davis, *Late Victorian Holocaust*, 388.

8. Duque, "Agricultura do Nordeste," 64.

9. Li, *Will to Improve*.

## Chapter One

1. Albuquerque, *Invenção do Nordeste*.

2. Metcalf, *Go-Betweens*.

3. Chandler, *Feitosas*, 129.

4. Brown, "Urban Growth."

5. Webb, *Changing Face*, 68; emphasis in the original.

6. Hall, *Drought and Irrigation*, 3.

7. Sobrinho, "Terra das Secas," 219.

8. Santos, *Cleansing Honor*, 33.

9. Chandler, *Feitosas*, 146.

10. Santos, *Cleansing Honor*, table 1.2.

11. Webb, *Changing Face*, 34.

12. Santos, *Cleansing Honor*, 133.

13. Swarnakar, "Drought, Misery and Migration," 5.

14. Cordeiro Manso, "A Seca no Ceará," undated ms., Biblioteca Nacional Music Collection, Rio de Janeiro, Brazil.

15. Cunniff, "Great Drought," 280.

16. Santos, *Cleansing Honor*, table 1.1.

17. Neves, "Seca na História," 83.

18. Andrade and Logatto, "Imagens da Seca."

19. Blake, *Vigorous Core*, 28.

20. Ibid., 46.

21. Santos, *Cleansing Honor*, 148.

22. Conselheiro adopted that name to distance himself symbolically from the earthly sociopolitical ties represented by his well-known family name, Maciel. Pessar, *From Fanatics to Folk*.

23. Levine, *Vale of Tears*, 89–106.

24. Borges, "Salvador's 1890s," 53.

25. Cunha, *Rebellion in the Backlands*.

26. Ibid., x–xviii.

27. Ibid., 106.

28. Stepan, *Hour of Eugenics*; Peard, *Race, Place, and Medicine*.

29. Levine, *Vale of Tears*, 6, 82, 110.

30. Rodrigues, *As collectividades anormaes*.

31. Pessar, *From Fanatics to Folk*, 28.

32. Blake, "Invention of the *Nordestino*," 97.

33. Greenfield, "*Sertão* and *sertanejo*," 37.

34. Cunha, *Rebellion in the Backlands*, 43.

35. Greenfield, "*Sertão* and *sertanejo*," 42.

36. Lyra, *As Secas do Nordeste*.

37. Roquette-Pinto, *Grêmio Euclides da Cunha*, 76.

38. Cunniff, "Great Drought," 186.

39. Studart, "Thomas Pompeu de Sousa Brasil," *Diccionário bio-bibliográphico*, 141–46.

40. Greenfield, *Realities of Images*, 89.

41. Raposo, Melo, and Andrade, *Firmino Leite*, 23–24.

42. Albuquerque, "Palavras que calcinam," 120.

43. Faoro, *Os donos do poder*, 620–54.

44. The republic's 1891 constitution granted suffrage to free adult males who could pass a modest literacy requirement—usually by signing their name. This was

an improvement over the previous criteria based on income, but it still justified the exclusion of a vast majority from the electorate, particularly in the rural northeast, where the literacy rate was lower than 10 percent (vs. 16 percent in Brazil as a whole). Lewin, *Politics and Parentela*, 23n36.

45. This "denunciation industry" is also suspect. Assistance to the northeast has routinely been criticized as wasteful, while lavish federal projects for other regions have not received equal scrutiny. Villa, *Vida e morte*, 252.

46. Aline Silva Lima, "Um Projeto," 26.

47. Hirschman, *Journeys toward Progress*, 23–25.

48. Ribeiro, *Brazilian People*, 245–46.

49. Ibid., 255.

50. Duarte, "Seca no Nordeste," 14.

51. Ribeiro, *Brazilian People*, 246.

52. Domingos and Hallewell, "Powerful"; Nelson and Finan, "Praying for Drought."

53. Souto, *Nordeste*; Bursztyn, *O Poder dos Donos*.

54. Mota, *Sertão alegre*, 92.

55. Slater, *Stories on a String*.

56. Campos, *Ideologia dos poetas populares*.

57. Curran, "Politics in the Brazilian *Literatura de Cordel*."

58. Slater, *Stories on a String*, 183.

59. Rubem Walter Moreira, "Flagelos da Seca no Nordeste," 1946, Cordéis, no. 2963, FCRB.

60. Leandro Gomes de Barros, "O Imposto e a Fome," Recife, 1909, Cordéis, no. 6054, FCRB.

61. Leandro Gomes de Barros, "A Sêcca do Ceará," Guarabira, Paraíba, 1920, Cordéis, no. 6063, FCRB.

62. Euclides da Cunha, "As Secas do Norte," *O estado de São Paulo*, October 29 and 30, November 1, 1900, reprinted in Cunha, *Canudos e ineditos*, 148–59.

63. Finan, "Climate Science," 208.

64. Davis, *Late Victorian Holocaust*.

## Chapter Two

1. Penna and Neiva, "Viagem científica," 165.

2. Blake, "Invention of the Nordestino," 89.

3. Farias, "Fase pioneiro," 71.

4. Fred L. Soper, "Survey of Pernambuco," submitted to director Wickliffe Rose for the 1920 International Health Board annual report, RG 5, series 2: IHB, subseries 305: Brazil, box 25, folder 150, 77–78, RAC.

5. "Relatório dos serviços realizados pela inspetoria de hygiene em 1912," *Archivos de Hygiene Pública e Medicina Tropical* 1, no. 1 (1915): 16.

6. Ibid., 141.

7. IHB personnel surveying disease incidence in the northeast also noted the migration of germs from coast to interior, but they blamed the Carnaval holiday and Catholic festivals rather than railway transportation for the human migrations that encouraged disease transmission. This assessment indicates their biases against rural folk culture. G. J. Carr on yellow fever in Parahyba do Norte [1925 or 1926], RG 5, series 2: IHB, subseries 305: Brazil, box 26, folder 156, RAC.

8. *Mensagem do Exm. Sr. Manoel Antonio Pereira Borba, Governador do Estado, lida por occasião da installação da 1a sessão da 9a Legislatura do Congresso Legislativo do Estado, aos 6 de março de 1916* (Recife: Imprensa Official, 1916), 27, FGE, APP.

9. *Mensagem do Exm. Sr. Dr. José Rufino Bezerra Cavalcanti, Governador do Estado, Lido ao installar-se a 2a sessão da 10a Legislatura do Congresso Legislativo de Pernambuco, aos 6 de Março de 1920* (Recife: Imprensa Official, 1920), 21, FGE, APP.

10. *Mensagem do Exm. Sr. Manoel Antonio Pereira Borba*, 40.

11. *Mensagem do Exm. Sr. Dr. José Rufino Bezerra Cavalcanti*, 5.

12. Sá, "Parahyba," 232.

13. Lewin, *Politics and parentela*, 23n36.

14. Studart, *Climatologia*, 12–14.

15. Ceará, *Relatório do inspector de hygiene pública, Dr. João Marinho de Andrade, ao exm. sr. presidente do estado*, May 1895, 443–46. Biblioteca Nacional.

16. Neto, *O poder e a peste*.

17. Teófilo, *Lyra rustica*, 41–51.

18. Teófilo, *A seca de 1915*.

19. Teófilo, *Seccas do Ceará*, preface.

20. Teófilo, *A fome*.

21. Ceará, *Relatório apresentado ao Exm. Sr. Dr. Aurelio de Lavor, secretário dos negócios do interior, pelo Dr. Carlos da Costa Ribeiro, inspetor de hygiene*, May 1916, 2–24. Biblioteca Nacional.

22. Rios, "O Curral."

23. *Relatório apresentado ao Exm. Snr. Dr. Aurelio de Lavor*, 39.

24. In 1906, Ceará's minister of interior commerce, a relative of Governor Accioly, accused Teófilo of having caused meningitis outbreaks through impure or improperly applied smallpox vaccine. The minister declared that only the government of Ceará should produce and distribute vaccine, to retain control over its quality and safety. Ceará, *Relatorio apresentado ao Exmo. Snr. Dr. Antonio Pinto Nogueira Accioly, presidente do estado do Ceará, pelo secretário de estado dos negócios do interior, José Pompeu Pinto Accioly*, May 1906, 73. Biblioteca Nacional.

25. Diniz, *O dinamismo*, 10–24.

26. Lutz and Machado, "Viagem Pelo Rio São Francisco"; Lima, "Missões Civilizatórias."

27. Peard, *Race, Place, and Medicine*.

28. Peixoto, *Clima e doenças*.

29. Peixoto, *Afranio vs. Afranio*.

30. Stepan, *Picturing Tropical Nature*.

31. Skidmore, *Black into White*, 74, 138.

32. Andrews, *Blacks and Whites*.

33. Foreign immigration to São Paulo was 18,000 annually in the 1880s but averaged 75,000 annually from 1891 to 1900. Katzman, "The Brazilian Frontier," 277; Dean, *Rio Claro*, discusses the protections offered by courts to disgruntled immigrants, which had not been available to slaves.

34. Lima and Britto, *Saúde e Nação*, 17.

35. Stepan, *"The Hour of Eugenics."*

36. A number of Nordestino politicians also advocated encouraging European immigration to their states, but these recommendations rarely led to concrete efforts. Blake, "The Invention of the Nordestino," chap. 2.

37. Bomfim, *A América Latina*; Skidmore, *Black into White*, 114–18.

38. These are collected in Torres, *A organização nacional*; and Torres, *O problema nacional*.

39. Skidmore, "Racial Ideas"; Skidmore, *Black into White*, 119.

40. McLain, "Alberto Torres," 17–34; Oliveira, *A questão nacional*, 122–24; Penna, *O Saneamento*, 71–76.

41. Simone Kropf argues persuasively that this was the date of the report's publication, despite the earlier date (1916) stamped on the volume. Kropf, *Doença de chagas*.

42. Penna and Neiva, "Viagem Cientifica," 199.

43. Ibid., 198.

44. Ibid., 172.

45. Ibid., 195.

46. Ibid., 223.

47. Ibid., 179.

48. Ibid., 170.

49. Ibid., 173.

50. Ibid., 183.

51. Ibid., 176.

52. The sertão had at times been viewed as a haven from diseases endemic on the coast due to its dry climate and sparse population. Studart, *Climatologia*, 6.

53. Penna and Neiva, "Viagem Científica," 161.

54. Ibid., 182.

55. Ibid., 174.

56. Pereira, "O Brasil." The nation-as-hospital metaphor reappeared subsequently in a variety of contexts.

57. Hochman, "Logo Ali," 221.

58. Ceará, *Relatório apresentado ao Exm. Sr. Dr. João Thomé de Saboya e Silva, M.D., presidente do estado, pelo secretário dos negócios do interior e da justiça, a Desmbargador José Moreira da Rocha*, June 1, 1919, 64. Biblioteca Nacional.

59. Meade, *Civilizing Rio*.

60. Needell, "The Revolta Contra Vacina."

61. Stepan, *The Beginnings of Brazilian Science.*

62. Lima, *Sertão chamado Brasil*; Diacon, *Stringing Together.*

63. Hochman, *A era do saneamento.*

64. Amado, "Região, sertão," 148.

65. Ibid., 148–49. The conceptual importance of the sertão in Brazilian state formation has been compared to the frontier in American history. Wagner, *A conquesta do oeste.*

66. Hochman, "Logo ali," 229.

67. Penna, *O saneamento*, 11.

68. Ibid., 172.

69. Ibid., 88.

70. Freitas, *Hygiene rural*, 8–15.

71. "Discurso do Dr. Acácio Pires na Paraíba," July 1921, BP/PI/TT/19210716, ABP COC.

72. Chagas replaced Cruz as director of Manguinhos (the Instituto Oswaldo Cruz) in 1917.

73. "Regulamento dos serviços do Departamento de Saúde Pública," *Revista Brasileira de engenharia* 1, no. 1 (1920): 44–46.

74. If the IHB was involved in the contract, it covered one-fourth of expenses and state obligations fell to one-fourth.

75. Penna, *Saneamento*, 164.

76. Federal funds for rural health services were, however, delayed for months, making programs inefficient and demoralizing staff. "Relatorio dos Serviços de Saneamento e Prophylaxia Rural," 1920, BP/PI/TP/19210408, ABP COC.

77. Ceará, *Relatório apresentado ao Exmo. Sr. Desembargador José Moreira da Rocha, secretário dos negócios do interior e justiça, pelo Dr. Carlos da Costa Ribeiro, diretor geral de hygiene,* May 1919, 57; emphasis added. Biblioteca Nacional.

78. Ibid., 6.

79. Castro Santos, "Power, Ideology."

80. Birn, *Marriage of Convenience*, table 5.2.

81. Campaigns against hookworm consisted of chenopodium treatments and installation of latrines to halt soil reinfection through human feces. Smillie, "The Results of Hookworm Disease Prophylaxis."

82. L. W. Hackett, "Notes on the Organization of the IHB Work in Brazil," April 28, 1919, RG 5, series 1.2, subseries 305, box 24, folder 145, RAC.

83. Wickliffe Rose to L. W. Hackett, December 13, 1920, RG 5, series 1.2, subseries 305, box 95, folder 1307, RAC.

84. L. W. Hackett to Wickliffe Rose, August 2, 1919, RG 5, series 1.2, subseries 305, box 78, folder 1110, RAC.

85. L. W Hackett to John A. Ferrell, January 5, 1921, RG 5, series 1.2, subseries 305, box 113, folder 1534, RAC.

86. "Plan and Progress of County Health Work in Brazil," 1923, RG 5, series 2, subseries 305, box 25, folder 154, RAC.

87. Farias, "A fase pioneiro," 72.

88. L. W. Hackett to Wickliffe Rose, November 28, 1919, RG 5, series 1.2, subseries 305, box 78, folder 1110, RAC.

89. L. W. Hackett to Wickliffe Rose, April 28, 1919, RG 5, series 1.2, subseries 305, box 78, folder 1109, RAC.

90. Paes de Azevedo, "Supplement to a Survey of Bahia," April 2, 1921, to L. W. Hackett, RG 5, series 2, subseries 305, box 25, folder 1150 RAC.

91. L. W. Hackett, "No. 7537 Report on Work for the Relief and Control of Hookworm Disease in Brazil from Nov. 22, 1916, to Dec. 31, 1919," December 24, 1920, RG 5, series 2, subseries 305, box 24, folder 148, RAC.

92. Fred Soper, "Survey of Pernambuco," 1920, RG 5, series 2, subseries 305, box 25, folder 1150, RAC.

93. Wickliffe Rose, "Observations on the Public Health Situation and Work of the IHB in Brazil," October 1920, RG 5, series 2, subseries 305, box 25, folder 153, No. 7502, 7–8, RAC.

94. Ibid., 177; emphasis added.

95. Lima, *Sertão chamado Brasil*, 104.


## Chapter Three

1. Alarico Araújo, telegram, March 6, 1920, Açude Russas file, 186.11, Fundo: Açudes Públicos, Arquivo da 1ª Diretoria Regional do DNOCS. (Hereafter, Arquivo DNOCS.)

2. Reis, *Relatório de obras novas*, 235.

3. Rodrigues, "As seccas."

4. Schwartzman, *Space for Science*, 68. Rio's Escola Politécnica was superseded as the nation's most prestigious engineering school when São Paulo's school of the same name was established in 1894.

5. "Estatutos do Clube de Engenharia," *Revista do clube de engenharia* 1 (1905): 151.

6. Carneiro, *Classe de engenharia civil*.

7. Costa, *Brazilian Empire*, 233.

8. Li, *Will to Improve*.

9. Lisboa probably drew this view from Euclides da Cunha's essentialist portrayal of sertanejo character and culture as having been molded by their harsh environment.

10. Lisboa, "O problema das seccas."

11. Ibid., 146.

12. Fournier, *O problema das secas*.

13. P. Florentino Barbosa, *O problema do norte*.

14. Walker, "Ceará (Brazil) Famines."

15. Ferraz, *Causes prováveis*; Ferraz, *A Previsão das seccas*.

16. Lisboa, "O problema das seccas," 140.

17. Proposals to divert some of the São Francisco River's flow to water other parts of the sertão remain volatile today. Rohter, "Brazilian Plan for Water Diversion."

18. GIFI files for the Ministério de Viação e Obras Públicas, 1912, 4B, box 283, AN.

19. Pessoa, *Obras do Nordeste*, includes Lisboa's response to an interview given by Pessoa.

20. M. A. Lisboa to MVOP, November 28, 2010, Açude Cedro file, 58.6, Fundo: Açudes Públicos, Arquivo DNOCS.

21. History of Cedro, 1909–1919 (reprints of telegrams), Açude Cedro file, 58.3, Fundo: Açudes Públicos, Arquivo DNOCS.

22. Relatório, 1914. Açude Cedro file, 58.9, Fundo: Açudes Públicos, Arquivo DNOCS.

23. Relatório, 1915. Açude Cedro file, 58.10, Fundo: Açudes Públicos, Arquivo DNOCS.

24. Relatório, 1917. Açude Cedro file, 58.4, Fundo: Açudes Públicos, Arquivo DNOCS.

25. Telegram, July 1919, Açude Cedro file, 58.10, Fundo: Açudes Públicos, Arquivo DNOCS.

26. Telegram from prefeito to drought inspector, January 1921, Açude Cedro file, 58.2, Fundo: Açudes Públicos, Arquivo DNOCS.

27. Letter, February 1920, Açude Cedro file, 58.2, Fundo: Açudes Públicos, Arquivo DNOCS.

28. Luetzelburg, *Estudo botânico*.

29. This and similar maps for other states in the drought zone are in the EP archive, Documentos Visuais, Aim 2, Pr4, Museu da República, Rio de Janeiro.

30. One acre-foot=43,560 cubic feet=325,900 gallons.

31. Waring, *Irrigation in Northeastern Brazil*, 8.

32. Ibid., 19.

33. Ibid., 7.

34. Sobrinho, "Açude Quixeramobim," 14, CPDOC, Ildephonso Simões Lopes Collection, ISL c 1910.08.25, file 5, FGV.

35. Ibid., 12.

36. This was due to both ecological factors (a leaf fungus that infected Amazon rubber) and economic competition from southeast Asian colonies. Dean, *Brazil and the Struggle for Rubber.*

37. Sobrinho, "Açude Quixeramobim," 225–36; this is similar to the draft at FGV.

38. Câmara Municipal de Quixeramobim, *Representação dirigida*, 9.

39. Neves, *O maior problema*, 23.

40. Silva, "O problema do norte," 13.

41. Barbosa, "Conferência realizada," 35–36.

42. Subsequent studies revealed that there was no rock base at the site sufficient to support the proposed dam, so construction at another site, Orós, was undertaken instead. Ribeiro, "Valle do Rio Salgado."

43. Bouchardet, *Seccas e irrigação*, 164–72.

44. Barbosa, "Conferência realizada," 7.

45. Sobrinho, "Açude Quixeramobim."

46. Ibid., 5–6.

47. The northeast's humanitarian crises paled in comparison to India's turn-of-the-century famines, in which millions died. Davis, *Late Victorian Holocaust*.

48. This perception was due to the Reclamation Service's emphasis on white farmer settlement in its promotional materials. The service also made limited attempts to turn Native Americans already occupying reclamation areas into irrigated smallholders.

49. Salgueiro, *Engenheiro Aarão Reis*, 30.

50. Upon being sent to oversee Obras Novas operations in Fortaleza, Reis wrote in his diary in September 1915, "I have never made a greater sacrifice in my professional life over more than forty years . . . I leave my darling old wife and my children apprehensive." Moraes, *O sertão descoberto*, 157.

51. Rodrigues, "As seccas do Ceará," 119.

52. Ibid., 110. In 1919 Brazil established forestry laws aimed at increasing plantings of Australian eucalyptus, a fast-growing, drought resistant species. IOCS operated two *hortos florestaes* (forest reserves) in Joazeiro, Bahia, and Quixadá, Ceará, from which many people obtained eucalyptus seeds. J. A. Fonseca Rodrigues recommended that Joazeiro trees, with their large canopies, and carnauba palms, for their marketable wax, also be planted in quantity.

53. Albano and Braga, *Obras de irrigação*.

54. Sobrinho, *Problema das seccas*.

55. It is probable that this is Paulo de Moraes Barros, appointed by Epitácio Pessoa in 1922 to a three-member commission that reviewed the progress of drought works during Pessoa's administration.

56. D. P. Robinson to inspector, August 16, 1923, Açude Orós file, 142.2, Fundo: Açudes Públicos, Arquivo DNOCS.

57. Mitchell, *Rule of Experts*.

58. Hardiman, "Politics of Water."

59. Whitcombe, "Environmental Costs of Irrigation." Stone, "Canal Irrigation," questions the significance of these problems.

60. In Brazil's northeast, a worm-borne malady known as *esquistossomose*, which primarily affected cattle, began spreading through canals following irrigation. Raposo and Melo, *Técnicos do DNOCS*, 109.

61. Worster, *Rivers of Empire*; Reisner, *Cadillac Desert*; Pisani, *Water and American Government*. Fiege, *Irrigated Eden*, observes that the intensity of irrigated farming requires farmers to have a vested interest in improving land productivity, which is less true when cultivators are merely tenants (as was often the case in the sertão).

62. Ervin, "1930 Agrarian Census."

63. Telegram, October 28, 1915, Açude Forquilha file, 88.4, Fundo: Açudes Públicos, Arquivo DNOCS.

64. Telegram, Forquilha to inspector, August 9, 1918, Açude Forquilha file, 88.5, Fundo: Açudes Públicos, Arquivo DNOCS.

65. Telegram, Abelardo dos Santos to inspector, July 1, 1919, Açude Forquilha file, 88.5, Fundo: Açudes Públicos, Arquivo DNOCS.

66. Telegram, Forquilha property owners to inspector, January 23, 1928, Açude Forquilha file, 88.6, Fundo: Açudes Públicos, Arquivo DNOCS.

67. Inspector to MVOP April 30, 1919. Fundo: Açudes Públicos, Açude Forquilha file, 88.5, Fundo: Açudes Públicos, Arquivo DNOCS.

68. In 1920, Brasil Sobrinho estimated that adequate daily food rations for adults during drought periods cost $350 réis; calorically adequate food rations for children cost $250 réis . Sobrinho, *O problema das seccas*, 28.

69. Abelardo dos Santos to inspector, August 7, 1919, Açude Forquilha file, 88.5, Fundo: Açudes Públicos, Arquivo DNOCS.

70. Abelardo dos Santos to inspector, September 26, 1919, Açude Forquilha file, 88.5, Fundo: Açudes Públicos, Arquivo DNOCS. Documents indicate that at least by January he had been permitted to do this (report dated April 10, 1920, same file).

71. Abelardo dos Santos to inspector, October 17, 1919, Açude Forquilha file, 88.5, Fundo: Açudes Públicos, Arquivo DNOCS.

72. In another instance, at the Saco Reservoir in Rio Grande do Norte, Reis fired an engineer for maintaining a sick worker on his work roles, although the engineer protested that he could not in good conscience leave the man and his family of six children to starve. Castro, *Avalanches*, 112.

73. Abelardo dos Santos to inspector, October 17, 1919; inspector reply, November 13, 1919, Açude Forquilha file, 88.5, Fundo: Açudes Públicos, Arquivo DNOCS.

74. Male workers occupied a range of salary tiers, from the engineer and his auxiliary to carpenters, electricians, guards, quarrymen, masons, mechanics, and blacksmiths, along with the vast army of manual laborers.

75. Abelardo dos Santos to inspector, April 10, 1920, Açude Forquilha file, 88.5, Fundo: Açudes Públicos, Arquivo DNOCS.

76. Castro, *Avalanches*, 91.

77. Lima, "Um projeto," 57.

78. During Epitácio Pessoa's presidency, when drought works were relatively well funded, farmworkers' wages across the northeast reportedly rose twofold to fourfold to compete with the many jobs offered at IFOCS construction sites. Lewin, *Politics and Parentela*, 330n44.

79. Lima, "Um projeto," chap. 3.

80. Ibid., 74–76, 95–97.

81. Epitácio Pessoa, correspondence with Governor João Pessoa, Archivo E. Pessoa, book 9, 129, 132, 134, IHGB.

82. Pessoa, *Pela verdade*, 335.

83. Ibid., 374.

84. "Falando à Nação," *Gazeta de Notícias* (Rio de Janeiro), August 5, 1923, 1.

85. Pessoa, *Pela verdade*, 351.

86. Ibid., 329.

87. Ministério de Viação e Obras Públicas, IFOCS Administração Central, 1ª seção, "Relatório apresentado ao Exmo. Sr. Inspetor Federal das Obras Contra as Secas, Dr. Miguel Arrojado Lisboa," by Ezequiel Ubatuba, Recife, Brazil, May 1920, CPDOC, Ildephonso Simoes Lopes Collection, ISL c 1910.08.25, file 5, FGV.

88. Albano and Braga, *Obras de irrigação*, 88.

89. Sobrinho, *Problema das seccas*, 13.

90. Albano, *O secular problema do Nordeste*, frontispiece.

91. The entire sertanejo population was estimated at 2.5 million in the early 1920s. Rondon, Barros, and Lopes, "Relatório apresentado."

92. Sobrinho, *Problema das seccas*, 59–60.

93. Ibid., 92.

94. Albano and Braga, *Obras de irrigação*, 125–29.

95. Ibid., 60.

96. Braga, *Seccas do Nordeste*.

97. Belissario Penna, "Brasil, futuro paraíso," undated manuscript [1919?], BP/PI/TP/90002040-49, ABP COC.

98. "Falando à nação," 2.

99. Pessoa, *Pela verdade*, 361.

100. Ibid., 335–36.

101. Rondon, a military engineer, studied at Brazil's Escola Superior de Guerra (military academy) under renowned positivist Benjamin Constant (Diacon, *Stringing Together*, 84–85). Lopes, a graduate of Rio's Escola Politécnica, was a federal representative from Rio Grande do Sul in 1906–8, 1913–19 and 1922–30 (*Dicionário histórico-biográfico* v. 3, 3270–72). Barros, a graduate of Rio's medical faculty, served as federal representative for São Paulo in 1909–11 and 1927–29 (*Dicionário histórico-biográfico* v. 1, 564–65).

102. Rondon, Barros, and Lopes, "Relatório apresentado," 48.

103. "Falando à nação," 1.

104. Rondon, Barros, and Lopes, "Relatório apresentado," 48.

105. Pessoa, *Obras completas*, 25:99–105; reprinted from *Jornal do Comércio* (Rio de Janeiro), September 12, 1933. The *Jornal* amended several of the commission's figures to be higher.

106. Rondon, Barros, and Lopes, "Obras do Nordeste," *Jornal do Comércio*, December 16–17, 1923, CPDOC, Ildephonso Simoes Lopes Collection, ISL, file 3 (draft in file 1), FGV.

107. Brasil, Comissão de Inspeção das Obras do Nordeste, *Obras do Nordeste*.

108. Pessoa, *Pela verdade*, 363.

109. Epitacio Pessoa response, December 27, 1923. CPDOC, Ildephonso Simoes Lopes Collection, ISL, file 3, FGV.

110. Sobrinho, "Capacidade irrigatória."

111. Pessoa, *Pela verdade*; see also Pessoa, *Obras do Nordeste*.

112. Veloso, *O Nordeste seco*.

113. Brandao, *Feixe de artigos*. The articles appeared in *Jornal do Comércio* and *O Norte*.

114. J. Palhano de Jesus, memo, Rio de Janeiro, October 23, 1929, CPDOC, Juarez Tavora archive, JT dt "Seca" folder, FGV.

115. On differential vulnerability to climate change see Endfield, *Climate and Society*.

116. On the concept of a technocratic "middle politics" see Ervin, "1930 Agrarian Census."

117. On rendering complex development challenges as technical problems see Li, *The Will to Improve*.

## Chapter Four

1. Almeida, *Trash/Bagaceira*, 14 and 16.

2. Da Cunha also expressed prejudice against coastal African or mixed-race Nordestinos as compared with hardier sertanejos. Da Cunha, *Rebellion in the Backlands*, 88; Blake, "Invention of the Nordestino," 84–85.

3. Campos, *Ideologia dos poetas*, 45–50.

4. Almeida, *Trash/Bagaceira*, 153. The translator uses the word "highlands" in place of *sertão*, drawing on a geographic and cultural distinction (highlander vs. lowlander) that he believed would be familiar to Anglophone readers.

5. The term "culture managers" comes from Williams, *Culture Wars*.

6. Dávila, *Diploma of Whiteness*.

7. Blake, "The Invention of the Nordestino," 334.

8. Ibid., 310–11; Sobrinho, "Terra das secas;" Sobrinho, "O homem do nordeste."

9. Weinstein, *For Social Peace in Brazil*.

10. B. Penna, "Porque sou integralista," June 29, 1937, BP/PI/TP/19370629, ABP COC.

11. Freyre, *Nordeste*; Coutinho, *O valor social da alimentação*.

12. Camargo, Raposso, and Flaksman, *O Nordeste*, 230.

13. Levine, *Father of the Poor?*

14. GV rc 940.10.30, Coleção Getúlio Vargas, Museu da República, Rio de Janeiro (news clips from Vargas's 1940 visit to the northeast compiled by the Departamento da Imprensa e Propaganda).

15. "Discurso pronunciado pelo Sr. Dr. Getúlio Vargas, chefe do governo provisório, no banquete realizado na cidade de João Pessoa, capital da paraíba do norte," *Jornal do Commercio* (Rio de Janeiro), September 9, 1933, Fundo Presidente Epitácio Pessoa, bk. 61, fols. 536–615, IHGB.

16. CODES 35 SDE, Presidente da República 1930–1945, Ministério de Viação e Obras Públicas, boxes 48–50 and 127–31, AN.

17. CODES 35 SDE, box 131, AN.

18. Silva, "Geologia do estado," 14.

19. Rodolfo Coelho Cavalcante, "O que Getulio Vargas fez pelo Brasil," n.d. [1940s], Cordel Collection no. 7392, FCRB.

20. Curran, "Politics in the Brazilian *Literatura de Cordel*," 118.

21. Later Almeida had a falling out with the Pessoa family over accusations that he gave preferential treatment to other friends in his political appointments. Arquivo Epitácio Pessoa, bk. 67, Fundo Presidente Epitácio Pessoa: José Américo Almeida, fols. 30–33, 42–62, 76–78, IHGB. See also Camargo, Raposso, and Flaksman, *O Nordeste*, 205.

22. These career details can be found in Almeida, *A Paraiba e seus problemas*, introduction and 723–28.

23. Almeida, *A Paraiba e seus problemas*, 593–637.

24. Ibid., vii.

25. "IFOCS Regulamento 1931," 13–16, CPDOC, FGV.

26. Vieira, "Obras no Nordeste."

27. "IFOCS até o fim de 1943," CPDOC, GV Archive, GV rem.s 1944.00.00, FGV.

28. Sobrinho, "Povoamento do Nordeste."

29. Frederico de Castro Neves, "A seca na história"; Kenny, "Landscapes."

30. Raposo, Melo, and Andrade, *Firmino Leite*, 10–11.

31. Camargo, Raposso, and Flaksman, *O Nordeste*, 211–230, 430.

32. Vieira, "O Ministro José Américo," 1.

33. Vieira, "Obras contra as seccas," 358.

34. The inspector was expected to travel to the northeast at least three times per year. "IFOCS Regulamento 1931," arts. 31 and 37, 16, CPDOC, FGV.

35. Raposo, Melo, and Andrade, *Firmino Leite*, 18.

36. Vieira, "Obras contra as seccas," 358.

37. Vieira, "As dificuldades," 176–79.

38. Vieira, "Obras contra as seccas," 357.

39. Before 1930, IFOCS had stored 621 million cubic meters of water in public reservoirs; that figure stood at 2.6 billion cubic meters following Vieira's first term. Ninety-three percent of the private reservoir capacity available in 1943 had been constructed under Vieira's direction. Almost 60 percent of the wells drilled by IFOCS through 1943 were drilled while Vieira was inspector, and he was responsible for nearly 75 percent of the highways constructed by the agency up to that time. "IFOCS até o fim de 1943."

40. Magalhães, "Drought and Policy Responses," 193.

41. "IFOCS até o fim de 1943."

42. Brazil, Ministry of Transportation and Public Works, *Service of Drought Control*, fig. 10.

43. Letter from engineers of IFOCS to ISL, April 12, 1937, CPDOC, Ildephonso Simoes Lopes Archives, ISL c 1910.08.25, file 2, FGV.

44. "Seccas do NE do Brasil," n.d. [c. 1930], CPDOC, ISL c 1910.08.25, file 2, FGV.

45. Camargo, Raposso, and Flaksman, *O Nordeste*, 192.

46. Humberto R. de Andrade, "Necessidade da collaboração do agronomo nas obras de irrigação," CPDOC, Juarez Tavora Files, JT dt Seca, file 1, FGV.

47. "Separata do relatório apresentado pelo Major Juarez Tavora, em Maio de 1932, ao Sr. Pres. da Republica . . . . : O problema das secas," CPDOC, Juarez Tavora Files, JT dt Seca, file 1, FGV.

48. Brasil, Ministério de Viação e Obras Públicas, "O problema social," 52.

49. Almeida, *As secas do Nordeste*, 14.

50. "Legislação do DNOCS" (Rio de Janeiro: MVOP-DNOCS, 1951), 76, CPDOC, FGV.

51. Vieira, "Obras do Nordeste."

52. Andrade, *Ensino e desenvolvimento*, 52–54.

53. "Sôbre o Aproveitamnto das Terras Irrigáveis do Nordeste," *Boletim da inspectoria de sêcas* 14, no. 2 (1940): 211–15.

54. Rabelo, *De* experts a *"bodes expiatórios."*

55. CPDOC, GV Archives, GV c 1943.09.28, roll 7, p. 3, FGV.

56. Wahrlich, *Reforma administrativa*, 664–66.

57. Compare this to the shift toward mechanized agricultural production in the United States during World War I. Fitzgerald, *Every Farm a Factory*.

58. Guerra, *O instituto agronômico*.

59. McCook, "Promoting the 'Practical.'" McCook finds that Colombia, Cuba, Puerto Rico, and Venezuela supported more rigorous agricultural research during the 1920s than was achieved in the Brazilian sertão, but research in those countries declined during the depression. McCook, *States of Nature*.

60. Raposo and Melo, *Têcnicos do DNOCS*, 141.

61. Ibid., 145.

62. Guerra, *Instituto agronômico*, 13. The amount originally allocated for São Gonçalo in 1934 was 3,000 contos.

## Chapter Five

1. "Irrigação," *Diário carioca* (Rio de Janeiro), February 1, 1940.

2. Duque, "Agricultura do Nordeste," 63.

3. "Eng. J. A. Trinidade," obituary, *Boletim da inspetoria de sêcas* 15, no. 1 (1941), n.p.

4. Trinidade, "Os serviços agricolas."

5. Trinidade, "Os postos agrícolas."

6. Trinidade, "Os serviços agricolas," 43.

7. Trinidade, *Os postos agricolas*, 7.

8. Trinidade, "Os serviços agricolas," 43.

9. Duque, "O fomento da produção," photo caption, n.p. Similar ideologies are evident among government-employed *técnicos* in twentieth-century Colombia and Mexico. See Safford, *Ideal of the Practical*; and Wolfe, *Watering the Revolution*.

10. Trinidade, "Postos agrícolas."

11. Ibid., 106.

12. Ibid., 112.

13. "Postos agrícolas da Inspetoria de Secas," *O campo* (Rio de Janeiro), March 1941, 40.

14. Guerra, *O instituto agronômico*.

15. Wolfe, "Bringing the Revolution."

16. Lopes de Andrade, *Introdução à sociologia*, 185. The author postulated that this solidarity made the culture particularly difficult to change.

17. Sobrinho, "Homem do Nordeste."

18. Duque, "O fomento da produção," 155.

19. Duque, *Vantagens da seca*.

20. Duque, *Solo e agua no polígono*, 2nd ed.,14.

21. Hargreaves, *Dry Farming* (1957), 83.

22. Duque, *Solo e agua no polígono*, 2nd ed., 15.

23. Ibid., 12.

24. Ibid., 111.

25. Ibid., 210.

26. Ibid., 111.

27. The Serviço Especial de Saúde Pública (Public Health Special Service), funded by the United States and overseen by Columbia University sociologist Charles Wagley, monitored the health of Nordestinos who opted to harvest rubber. Garfield, *In Search of the Amazon*; Campos, "International Health Practices."

28. Guerra, *O instituo agronômico*, 44.

29. Duque, *Solo e agua no polígono*, 3rd ed., 7.

30. Ibid., 201 and 291.

31. Ibid., 188 and 201.

32. Ibid., 200.

33. In 1939 Rs12$500 was equal to US$1.00.

34. Barreira, "Observações," 13.

35. Nobrega, "Ensaio social-econômico," 14.

36. Notably, six of the eleven property owners that Nobrega interviewed were literate.

37. Nobrega, "Ensaio social-econômico," 10.

38. Ibid., 12.

39. Ibid., 14.

40. *Irrigante amigo! Seja bem vindo ao Projeto São Gonçalo*, undated pamphlet, Instituto Agronômico José Augusto Trinidade Collection, DNOCS library, Fortaleza.

41. Serviço Agro-Industrial, "Informações para irrigantes" (Fortaleza: Ministério de Viação e Obras Públicas/DNOCS, 1957), Instituto Agronômico José Augusto Trinidade Collection, DNOCS library, Fortaleza.

42. Tigre, *Catecismo*, 10.

43. Ibid., 12.

44. Ibid., 45.

45. Ibid., 51.

46. Castro, *Geografia da fome*.

47. Tigre, *Catecismo*, 54.

48. Gomes, "Solução agronômica."

49. Duque, "Agricultura do Nordeste," 52.

50. Ibid., 64.

51. Ervin, "1930 Agrarian Census."

52. Memo to Ministério dos Negócios do Trabalho, Indústria e Commércio, copied to Ministério de Viação e Obras Públicas/IFOCS, December 1939. Choró Reservoir files, file 4, Fundo: Açudes Públicos, Arquivo da 1ª Diretoria Regional do DNOCS. (Hereafter, Arquivo DNOCS.)

53. Memos responding to the memo cited in fn 52, 1940, Choró Reservoir files, file 4, Fundo: Açudes Públicos, Arquivo DNOCS.

54. DNOCS inspector Berredo to MVOP, 1948, Choró Reservoir files, file 1, Fundo: Açudes Públicos, Arquivo DNOCS.

55. Memo from José Nanges Campo discussing Delfino Alencar case, March 1948, Choró Reservoir files, file 1, Fundo: Açudes Públicos, Arquivo DNOCS.

56. Letters from J. G. Duque to drought inspector, June 20, 1944, Forquilha Reservoir files, file 7, Fundo: Açudes Públicos, Arquivo DNOCS.

57. Letters from J. G. Duque to drought inspector, November 1946, Forquilha Reservoir files, file 7, Fundo: Açudes Públicos, Arquivo DNOCS.

58. Seven memos from J. G. Duque to other DNOCS personnel, April 10, 1953–September 3, 1954, Forquilha Reservoir files, file 7, Fundo: Açudes Públicos, Arquivo DNOCS.

59. Letter from Associação do Commércio, Sobral, to DNOCS director, April 24, 1957, Forquilha Reservoir files, file 7, Fundo: Açudes Públicos, Arquivo DNOCS.

60. Caldas, "Questões de seccas."

61. Curran, "Politics in the Brazilian *Literatura de Cordel*."

62. J. Evilasio Tavares, "A viagem de um poeta atravéz das labias dos engenheiros das obras contra as seccas," João Pessoa, Pernambuco, Cordel no. 7055, FCRB.

63. Raimundo Santa Helena, "Flagelados das secas," Paraíba, 1932. Cordel no. 7867 (bound with his "Duelo do Padim Ciço com o Papa"), FCRB. *Angu* is a simple dish made from corn flour, water, and salt.

64. Pedro Brazil, "A ceca do Ceará," 1922, Cordel no. 3937, FCRB.

65. Manoel Camilo dos Santos, "Horrores do Nordeste e solidariedade campinense," Campina Grande, 1953, Cordel no. 1479, FCRB.

66. Rodolfo Coelho Cavalcante, "Os flagelos da sêca do Nordeste," Salvador (?), n.d., Cordel no. 4712, FCRB; Curran, *A presença de Rodolfo Coelho Cavalcante*, includes an anthology of Cavalcante's extensive corpus.

67. Padre Cícero claimed to have worked a miracle in 1889, turning communion wine into the physical blood of Christ. He inspired a great following, which continued beyond his death in 1934.

68. João Martins de Athayde, "Retirante," Recife, 1946, Cordel no. 2462, FCRB; José Bernardo da Silva, "Retirante" [possibly a copy of Athayde's?], Juazeiro, Ceará, 1955, Cordel no. 1254, FCRB.

69. Severino Borges Silva, "Sêca! fome! e carestia! assola o (Brasil) inteiro," n.d., n.p. [Recife, 1953?], Cordel no. 1235, FCRB.

70. Manoel Domingues Ferreira, "Catastrophe do Nordeste," n.d. [1960?], Cordel no. 4945, FCRB.

71. See Williams, *Culture Wars in Brazil*, esp. chap.3.

72. McCann, *Hello, Hello Brazil*, chap. 3.

73. Ibid., 119.

74. Luiz Gonzalga, "Asa branca," lyrics by Humberto Teixeira, music by Luiz Gonzaga, RCA-Victor, 1947.

75. Luiz Gonzalga, "A volta da asa branca," lyrics by Zé Dantas, music by Luiz Gonzaga, RCA-Victor, 1949.

76. Vieira, *O sertão em movimento*, 173–75.

77. João Martins de Atahyde, "Suspiros de um sertanejo," Recife, 1945, Cordel no. 1289, FCRB.

78. Pedro Alves da Silva, "Poesias do sertão," 1937, Cordel no. 1199, FCRB.

79. Luiz Gonzaga, "Vozes da seca," lyrics by Zé Dantas, music by Luiz Gonzaga, RCA-LEME, 1953. Citations to Gonzaga's drought-related songs are from Phaelante, "A seca do Nordeste."

80. Luiz Gonzaga, "Paulo Afonso," lyrics by Zé Dantas and Dilo Melo, music by Luiz Gonzaga, RCA-Victor, 1955. There is another *baião* by the same name that hails the new plant for illuminating the whole northeast.

81. Pisani, *Water and American Government*, chap. 9; Hargreaves, "Introduction," in *Dry Farming: Years of Readjustment*.

82. Neves, "Getúlio e a seca," 127.

83. Author's visit to Russas, Ceará, irrigation perimeter in July 2010, guided by agronomist Vandemberk Rocha de Oliveira.


Chapter Six

1. Kirkendall, *Social Scientists*, sees economists as among a new group of "service intellectuals" who applied their expertise to social problems.

2. Of course, progressive economists in the United States also met their share of resistance from influential business leaders—especially following the New Deal. Kirkendall, *Social Scientists*.

3. Thomaz Pompeu de Souza Brasil Sobrinho estimated the 1935 sertão population at under three million, though the territorial boundaries he used may have been smaller than those employed by the census takers—and his estimates are probably less reliable. Sobrinho, "Povoamento do Nordeste."

4. *Legislação do DNOCS*, population tables at end, n.p.

5. Vinícius Berredo, *Obras contra as secas*, Conference presentation, February 8, 1950, Ministério de Viação e Obras Públicas-DNOCS, DNOCs library.

6. Duque, *Ligeiro estudo*.

7. Baity, Albuquerque, and Branco, "Estudos preliminares."

8. Serebrenick, "Comissão do vale"; Robock, *Brazil's Developing Northeast*, 86.

9. Callado, *Sêca fria*; Lima, *Terra nordestina*.

10. Motta, *Como resolver*.

11. Almeida, *As secas do Nordeste*, 9–10.

12. A. da Silva Mello, *Nordeste Brasileiro*, sec. 1, chap. 7.

13. Vargas, *Presidential message*, 1.

14. Robock, *Brazil's Developing Northeast*, 88.

15. Banco do Nordeste do Brasil, *Planejamento do combate*, 18.

16. Galvão, *O desenvolvimento económico*, 11.

17. Peterson, *Pesquisa econômica*, 5.

18. Robock, "The Negro in the Industrial Development of the South."

19. Robock, *Brazil's Developing Northeast*, 136.

20. Campos, *Realidade económica*.

21. Hillman, *O desenvolvimento economico*. Hillman was an economist from the Brazilian-U.S. agricultural office of the American embassy, part of the Point IV mission.

22. Robock, *Brazil's Developing Northeast*, 8.

23. Alexander, *Juscelino Kubitschek*, 179.

24. Roett, *Politics of Foreign Aid*, 53.

25. Grupo de Trabalalho para o Desenvolvimento do Nordeste, *Uma política de desenvolvimento*, 70.

26. Castro, *Death in the Northeast*, 136, quoting his 1956 speech before the Federal Chamber of Deputies.

27. Grupo de Trabalho para o Desenvolvimento do Nordeste, *Política de desenvolvimento*, 91.

28. Meira, *A luta contra as secas*, 74.

29. Raposo, Melo, and Andrade, *Firmino Leite*, 27.

30. Raposo and Melo, *Técnicos do DNOCS*, 90–91, 94–96.

31. "Omissão do Banco do Nordeste," *Jornal do Commércio* (Recife), April 6, 1958.

32. "'Piores que a seca, no nordeste, são o latifundiarismo e o feudalismo agrario,'" *Folha da manhã* (São Paulo), May 21, 1958. One source of inefficiency in Northeastern agricultural production was U.S. price guarantees for Brazilian sugar. See Levinson, *Alliance That Lost Its Way*, 244.

33. Furtado, "Policy," 3.

34. Ibid., 6.

35. Furtado, *A operação Nordeste*.

36. Manuel Correia de Andrade, *Geografia economica*, 84–88.

37. Furtado, *A operação Nordeste*, 63.

38. Furtado, *A fantasia desfeita*, 65.

39. Andrade, *A reforma agrária*, 12.

40. For text of the proposed law in its original and revised forms, see Andrade, *A reforma agrária*, 59–72.

41. Memo from DNOCS director José Cândido Castro Parente Pessoa to MVOP Ernani do Amaral Peixoto, October 26, 1959, collection EAP 59.08.01 vop, file 2, second folder, FGV.

42. Castro, *Death in the Northeast*, 20.

43. This view seems to dismiss the Marxist inspiration of urban northeastern reformers in the 1950s, such as Paulo Freire, who may have indirectly influenced formation of the Peasant Leagues. Kirkendall, *Paulo Freire*.

44. Furtado, *Fantasia desfeita*, 67.

45. See the articles collected in Callado, *Sêca fria*.

46. Callado, *Os industriais da sêca*, 54. The articles comprising this volume were published September–December 1959.

47. Ibid., 57.

48. Memo from José Candido Castro Parente Pessoa to MVOP Ernani do Amaral Peixoto, September 21, 1959. Collection EAP 59.08.01 vop, file 2, second folder, FGV; Despachos, October 9 and 26, 1959, Collection EAP 59.08.01 vop, file 2, second folder, FGV.

49. Callado, *Os industriais da sêca*, 77.

50. Wright and Wolford, *To Inherit the Earth*, 127–30.

51. Furtado, *Fantasia desfeita*, 80.

52. Celso Furtado, "Porque *Sudene?*," February 1962, Superintendência de Desenvolvimento do Nordeste library.

53. Roett, *Politics of Foreign Aid*, 40–41.

54. Superintendência de Desenvolvimento do Nordeste, *A Policy for the Economic Development of the Northeast*, 47.

55. Brasil, Superintendência de Desenvolvimento do Nordeste, *I plano diretor*, 151.

56. Furtado, *Fantasia desfeita*, 86.

57. José Candido Castro Parente Pessoa to MVOP Ernani do Amaral Peixoto, Despacho no. 5, June 20, 1960, Collection EAP 59.08.01 vop, file 2, second folder, FGV.

58. Ibid.

59. Furtado, *Fantasia desfeita*, 84–85.

60. Acervo EAP, file 1, file titled Açude de Orós, "Discurso pronunciado na inauguração do açude de Orós," FGV.

61. *Orós: Açude de esperança*, 1960, Departamento Nacional de Obras Contra as Secas library.

62. For examples, see Rodolfo Coelho Cavalcante, "Inundação do Rio São Francisco em 1946," Joazeiro, Bahia, Cordel no. 2285, FCRB; Manoel A. Campina, "Cheia de 48," Cordel no. 1155, FCRB; Moisés Matias de Moura, "Inundação do 5 maio 1949," Cordel no. 1219, FCRB; Alberto Porfirio Silva, "O Arrombamento do Orós," Cordel no. 3964, FCRB; Joaquim Batista de Sena, "História das inundações do Orós e as vítimas do Vale do Jaguaribe," Fortaleza, Cordel no. 6013, FCRB; and Maoel Domingues Ferreira, "Catástrophe do Nordeste," Cordel no. 4945, FCRB.

63. Robock, *Brazil's Developing Northeast*, 111.

64. Ibid., 173.

65. Robock, "What the United States Can Learn."

66. Antonio Rodrigues Coutinho, *A estrutura agrária Brasileira*.

67. Albuquerque, *Desenvolvimento economico*.

68. Page, *Revolution That Never Was*, 132–40.

69. Roett, *Politics of Foreign Aid*, 95–96.

70. Ibid., xi.

71. Miguel Arraes, interview, *Newsweek*, March 11, 1963, 55.

72. Skidmore, *Politics in Brazil*, 289.

73. Page, *Revolution That Never Was*, 208.

74. Ibid., 217.

75. "Sudene," Pronumerário no. 0276B, archive 1, Fundo SSP n. 444, DOPS (Delegacia de Ordem Política e Social) collection, Acervo Público de Pernambuco.

76. Furtado, *Fantasia desfeita*.

## Conclusion

1. Duque, "As formas de assistencia," 345.

2. Carvalho, *A economia política*, 233–51.

3. Ministério do Interior/José Osvaldo Pontes, "O trabalho do DNOCS e o programa de irrigação no Nordeste semi-árido," lecture presented to the Comissão de Agricultura do Senado Federal (Agricultural Commission of the Federal Senate), May 1978, Biblioteca do Senado.

4. Carneiro and Monte-Mór, "Sujeição e idealização."

5. MINTER/DNOCS, Divisao de Assistencia as Comunidades, "Manual de operacionalização dos trabalhos de promocão social e cooperativista," 1976, DNOCS library, Fortaleza.

6. Gomes, "A política de irrigação."

7. Hall, *Drought and Irrigation*, 109–25.

8. Oliveira, *Elegia*.

9. José Osvaldo Pontes, "O DNOCS e a irrigação do Nordeste," paper presented at the conference III Seminário Nacional de Irrigação e Drenagem, November 18, 1975, 8–10, DNOCS library, Fortaleza.

10. Filho, *As secas do Nordeste*. Similarly, the Serviço Especial de Saúde Pública (Special Public Health Service) claimed to have greatest success in the 1950s when it implemented low-tech sanitation technologies, like pit latrines, over more expensive waste disposal systems. Milanez, "Das atividades," 330.

11. Pereira, *Aspectos economicos e sociais*, 54–55.

12. Andrade, *Geografia economica*, 18.

13. Andrade, *Geografia, sociedade, e cultura*, 77–79, 85–89.

14. Furtado, *O Nordeste*, 17.

15. Furtado, *Seca e poder*.

16. Bond, "A Drought Ravages Northeast Brazil."

17. Nelson and Finan, "Praying for Drought."

18. Tendler, *Good Government*, chap. 3.

19. Silva, "Entre dois paradigmas."

20. Magalhães and Magee, "The Brazilian Northeast."

21. Departamento Nacional de Obras Contra as Secas, *100 Anos.*

22. Compare the analysis of the U. S. Bureau of Reclamation in Pisani, *Water and American Government,* with that in Worster, *Rivers of Empire.*

23. Meade, *Civilizing Rio.*

# Bibliography

## Archival Sources

*Brazil*

Brasília, Federal District
    Biblioteca do Senado (Federal Senate Library)
Fortaleza, Ceará
    Banco do Nordeste do Brasil, library
    Departamento Nacional de Obras Contra as Secas, library and archive
        Açudes Públicos, Arquivo da 1ª Diretoria Regional do DNOCS
Recife, Pernambuco
    Acervo Público de Pernambuco, state archive
        Acervo Delegacia de Ordem Política e Social (DOPS)
        Fundo Governador do Estado
    Centro Josué de Castro, archive
        Coleção Recortes
    Fundação Joaquim Nabuco, archive and library for Nordeste culture
    Superintendência de Desenvolvimento do Nordeste, library
    Universidade Federal de Pernambuco, library
Rio de Janeiro
    Academia Nacional de Medicina, academic library
    Arquivo Nacional
    Biblioteca Edison Carneiro, Museu de Folclore
    Biblioteca Nacional
        Música e Arquivo Sonoro
        Publicações Seriadas
    Casa de Oswaldo Cruz, Fundação Oswaldo Cruz, public health archive and
        library
        Arquivo Belisário Penna
    Centro de Pesquisa e Documentação de História Contemporânea do Brasil,
        Fundação Getúlio Vargas, archive for post-1930 era
        Arquivo Ernani do Amaral Peixoto
        Arquivo Getúlio Vargas
        Arquivo Ildephonso Simões Lopes
        Arquivo Juarez Tavora
        História Oral
    Clube de Engenharia, archive and library

Fundação Casa de Rui Barbosa, archive and library for First Republic period
  Acervo Cordel
Instituto Histórico-Geográfico Brasileiro, academic library
  Fundo Presidente Epitácio Pessoa
Museu da República
  Coleção Epitácio Pessoa, Documentos Visuais
  Coleção Getúlio Vargas

*United States*

Tarrytown, NY
  Rockefeller Archive Center
    RG5, series 2: IHB, subseries 305: Brazil
Washington, DC
  Library of Congress
  Smithsonian Institution Archives

## Interviews Conducted

Agronomists at Russas, Ceará, 2010.
Engineers at the Departamento Nacional de Obras Contra as Secas and the
  Federal University of Ceará in Fortaleza, 2002.

## Periodicals Surveyed

*Archivos de Hygiene Pública e Medicina Tropical*, Pernambuco
*Boletim da Inspetoria Federal de Obras Contra as Secas*, Ministério da Viação e
  Obras Públicas, Rio de Janeiro
*Hygia*, Rio Grande do Sul
*Prophylaxia*, Rio de Janeiro
*Revista Brasileira de Engenharia*, Rio de Janeiro
*Revista Brasileira de Geografia*, Rio de Janeiro
*Revista do Clube de Engenharia*, Rio de Janeiro
*Revista do Instituto Archeologico e Geográfico Pernambucano*, Recife
*Revista do Instituto do Ceará*, Fortaleza
*Revista do Serviço Especial de Saude Pública*, Rio de Janeiro
*Saneamento*, journal of the Serviço de Saneamento Rural, Rio de Janeiro

## Published Primary Sources

**Note**: The list of published primary sources below does not include all individual
government reports cited in the text.

Albano, Ildefonso. *O secular problema do Nordeste: Discurso pronunciado na
  câmara dos deputados em 15 de Outubro de 1917*. Rio de Janeiro, 1917.

Albano, Ildefonso, and Cincinato Braga. *Obras de irrigação para o Nordeste num ante projeto de 1919.* Mossoró, Rio Grande do Norte: Fundação Guimarães Duque, 1988.

Albuquerque, Luis Saboya de. *Desenvolvimento economico e social dos vales beneficiados por obras do DNOCS.* Brasília: DFP/DAE, 1964.

Almeida, José Américo de. "O Ministro José Américo e o Nordeste." *A União,* August 3, 1934, 13–14.

——. *A Paraiba e seus problemas,* 4th ed. Brasilia: Senado Federal, 1994.

——. *As secas do Nordeste,* 2nd ed. Vol. 177 of *Coleção Mossoroense.* Mossoró, Rio Grande do Norte: Fundação Guimarães Duque, 1981.

——. *Trash/Bagaceira.* Translated by R. L. Scott-Buccleuch. London: Owen, 1978.

Andrade, F. Alves de. *A reforma agrária no polígono das sêcas.* Fortaleza: Imprensa Oficial do Ceará, 1959.

Andrade, Lopes de. *Introdução à sociologia das sêcas.* Rio de Janeiro: Editôra a Noite, 1948.

Andrade, Manuel Correia de. *Geografia economica do Nordeste.* São Paulo: Editora Atlas, 1970.

Arraes, Miguel. Interview. *Newsweek,* March 11, 1963, 55.

Baity, Herman G., Manuel H. Barbosa de Albuquerque, and Carlos Couto Castello Branco. "Estudos preliminares para as obras sanitarias no Vale do São Francisco." *Revista do Serviço Especial de Saúde Pública* 4, no. 3 (1951): 671–945.

Banco do Nordeste do Brasil. *Planejamento do combate às secas.* Publication no. 4. Fortaleza: Banco do Nordeste do Brasil, 1953.

Barbosa, J. C. de Castro. "Conferência realizada em sessão do Conselho Director, 16 de maio 1908." *Revista do Clube de Engenharia* 22 (1910): 35–36.

Barbosa, P. Florentino. *O Problema do Norte.* Parahyba do Norte, Paraíba: Typografia da Livraria Gonsalves Penna, 1913.

Barreira, Inacio Ellery. "Observações sôbre as condições de vida e de trabalho do operário rural." *Boletim da Inspetoria Federal de Obras Contra as Sêcas* 17, no. 1 (1942): 12–14.

Berredo, Vinícius. *Obras contra as secas: Conferência realizada 8 Fevereiro 1950.* Fortaleza: MVOP-DNOCS.

Bomfim, Manoel. *A América Latina: Males de origem.* Rio de Janeiro: H. Garnier, 1905.

Bouchardet, Joanny. *Seccas e irrigação: Solução scientífica e radical do problema nordestina, geralmente intitulado o problema do Norte.* Rio Branco, Minas Gerais, 1938.

Braga, Cincinato. *Seccas do Nordeste e reorganização económica.* Rio de Janeiro, 1919.

Brandão, E. Souza, *Feixe de artigos: Contribuições para a minoração dos effeitos que as seccas motivam em nossa região Nordeste.* 2nd ed. Mossoró, Rio Grande do Norte: Fundação Guimarães Duque, Coleção Mossoroense, 1987. Originally published 1920.

Brasil, Comissão de Inspeção das Obras do Nordeste, General Cândido Rondon, Paulo de Moraes Barros, and I. Simões Lopes. *Obras do Nordeste: Resposta da Commissão de Inspecção ao Dr. Epitácio Pessôa*, 1924.

Brasil, Departamento Nacional de Obras Contra as Secas. *100 anos de atuação no estado do Ceará*. Fortaleza: INESP, 2010.

Brasil, Ministerio da Viação e Obras Públicas. *O problema social e econômico das obras contra as secas: Relatorio apresentado em Julho 1933 ao Snr. Minístro da Viação e Obras Públicas, Dr. José Américo de Almeida, pelo Eng. Agr. Evaristo Leitão*. Rio de Janeiro: Ministerio da Viação e Obras Públicas, 1937.

Brazil, Ministry of Transportation and Public Works, Federal Department of Drought Control Service. *Service of Drought Control in the Brazilian Northeast*. Rio de Janeiro: Ministry of Transport and Public Works, 1939.

Brasil, Superintendência de Desenvolvimento do Nordeste. *I plano director 1961–63*. Recife: SUDENE, 1966.

Caldas, L. Raul de Sena. "Questões de seccas." *Revista Brasileira de Engenharia* 8, no. 5 (1924): 224–29, and 8, no. 6 (1924): 247–50.

Callado, Antonio. *Os indústriais da sêca e os "Galileus" de Pernambuco: Aspectos da luta pela reforma agrária no Brasil*. Rio de Janeiro: Editôra Civilização Brasileira, 1960.

———. *A sêca fria*. Rio de Janeiro: Ministério de Educação e Cultura, 1961.

Câmara Municipal de Quixeramobim. *Representação dirigida ao Exmo. Snr. Dr. J. J. Seabra, Minístro da Viação e Obras Públicas*. Fortaleza: Typografia Escolar, 1911.

Camargo, Aspasia, Eduardo Raposso, and Sergio Flaksman. *O Nordeste e a política: Diálogo com José Américo de José Américo*. Rio de Janeiro: Nova Fronteira, 1984.

Carneiro, B. Piquet. *A classe de engenharia civil no regimen Republicano e a situação política do paiz*. Rio de Janeiro: Typografia Revista dos Tribunaes, 1921.

Carneiro, Maria José, and Patricia Monte-Mór. "Sujeição e idealização do passado: Reflexões sobre as representações das condições de vida do 'irrigante,'" *Reforma Agrária* 13, no. 3 (1983): 27–36.

Campos, Aluízio Affons. *Realidade econômica e planejamento do Nordeste*. ETENE publication no. 4. Fortaleza: Banco do Nordeste do Brasil, 1956.

Castro, Josué de. *Death in the Northeast*. New York: Random House, 1966.

———. *A Geografia da fome*. Rio de Janeiro: Editora O Cruzeiro, 1946.

Cavalcanti, Plinio. "A Canaan sertaneja." *Saúde* 1, nos. 4–6 (1918): 265–81.

Chagas, Carlos. "Moléstia de Carlos Chagas." *Brasil Médico* 25 (1911): 340–43, 353–55, 361–64, 373–75.

Chaves, Nelson. *O problema alimentar do Nordeste Brasileiro: Introdução ao seu estudo econômico social*. Recife: Editora Medico Cientifica, 1946.

Chaves, Nelson. *A Sub-alimentação no Nordeste Brasileiro*. Recife: Imprensa Oficial, 1948.

Coutinho, Antonio Rodrigues. *A estrutura agrária Brasileira nos censos de 1950 e 1960*. Brasília: Ministério da Agricultura, Departamento Econômico, 1963.

Coutinho, Ruy. *O valor social da alimentação*. 2nd ed. Rio de Janeiro: Agir Editora, 1947.

Crandall, Roderic. *Geographia, geologia, supprimento d'agua, transportes e açudagem nos estados da Parahyba, Rio Grande do Norte e Ceará*. Inspetoria Federal de Obras Contra as Secas, publication no. 4, 1910.

Cunha, Euclides da. *Canudos e ineditos*. São Paulo: Edições Melhoramentos, 1967.

———. *Rebellion in the Backlands*. Translated by Samuel Putnam. Chicago: University of Chicago Press, 1944. Originally published as *Os Sertões*, 1902.

Dias, João de Deus de Oliveira. *O problema social das secas em Pernambuco*. Recife, 1949.

Duque, J. G. "Agricultura do Nordeste e o desenvolvimento econômico." *Boletim do Departamento Nacional de Obras Contra as Secas* 19, no. 4 (1959): 48–72.

———. "O fomento da produção agrícola." *Boletim da Inspetoria Federal de Obras Contra as Sêcas* 11, no. 2 (1939): 150–64.

———. "As formas de assistencia aos irrigantes." *Boletim do Departamento Nacional de Obras Contra as Secas* 23, nos. 13–14 (1965): 343–49.

———. *Ligeiro estudo sobre irrigação no Nordeste*. ETENE publication no. 12. Fortaleza: Banco do Nordeste do Brasil, 1959.

———. *Melhoramento dos pastos do Nordeste*. Fortaleza: Companhia Cearense de Desenvolvimento Agropecuário, 1967.

———. *Solo e agua no polígono das sêcas*. 2nd ed. Fortaleza: MVOP-DNOCS, 1951.

———. *Solo e agua no polígono das sêcas*. 3rd ed. Fortaleza: MVOP-DNOCS, 1953.

———. *Vantagens da seca*. Coleçao Mossoroense series B, no. 509. Mossoró, Rio Grande do Norte: Fundação Guimarães Duque, 1988.

"Falando à nação." *Gazeta de Notícias*. Rio de Janeiro August 5, 1923, 1–2.

Ferraz, J. de Sampaio. *Causas prováveis das sêccas do Nordeste Brasileiro: Conf. realisada no Club de Engenharia no dia 20 de Dezembro de 1924*. Rio de Janeiro: Ministério da Agricultura, Indústria e Commércio, 1925.

———. "Contingente meteorológico para solução do problema das seccas." *Revista do Clube de Engenharia* 33 (1931): 217–27.

———. *A previsão das seccas do Nordeste: Ensaios pelo méthodo de correlações*. Rio de Janeiro: Ministério da Agricultura, Indústria e Commércio, 1929.

Fielo, Mariano. "Perspectivas da açudagem no Nordeste seco." *Revista Brasileira de Geografia* 16, no. 2 (1954): 213–27.

Fournier, Luiz Mariano de Barros. *O problema das secas do Nordeste*. 2nd ed. Series C, vol. 475 of *Coleção Mossoroense*. Mossoró, Rio Grande do Norte: Fundação Guimarães Duque, 1989.

Freitas, Octavio de. *Hygiene rural*. Recife: Imprensa Industrial, 1918.

Freyre, Gilberto. *Açúcar: Em tôrno da etnografia, da história, e da sociologia do doce no Nordeste canavieiro do Brasil*. 2nd ed. Rio de Janeiro: Ministério da Indústria e do Comércio, Instituto do Açúcar e do Alcool, 1969.

———. *Manifesto regionalista*. Recife: Instituto Joaquim Nabuco, 1976.

———. *Nordeste*. Rio de Janeiro: Livraria José Olympio, 1937.

Furtado, Celso. *A fantasia desfeita*. Rio de Janeiro: Paz e Terra, 1989.

——. *A operação Nordeste*. Rio de Janeiro: Ministério de Educação e Cultura, 1959.

——. "A Policy for the Economic Development of the Northeast" (1958). Reprinted in Superintendência de Desenvolvimento do Nordeste, *A Policy for the Economic Development of the Northeast and a Synthesis of the First Guiding Plan for the Development of the Northeast*, 15–79. Recife: SUDENE, 1961.

Galvão, Olavo João. *O desenvolvimento económico do Nordeste: Ausencia de coordenação regional*. ETENE publication no. 9. Fortaleza: Banco do Nordeste do Brasil, 1956.

Gomes, Gustavo Maia. "A politica de irrigacão no Nordeste: Intenções e resultados." *Pesquisa e Planejamento Economico* 9, no. 2 (1979): 411–45.

Gomes, Pimentel. "Solução agronômica do problema das sêcas." *Boletim do Departamento Nacional de Obras Contra as Secas* 19, no. 3 (1959): 113–24.

Grupo de Trabalho para o Desenvolvimento do Nordeste. *Uma política de desenvolvimento para o Nordeste*. 2nd ed. Recife: Ministério do Interior, SUDENE, 1967.

Guerra, Paulo de Brito. *O instituto agronômico José Augusto Trindade*. Fortaleza: Ministério do Interior, DNOCS, 1984.

Hillman, Jimmy S. *O desenvolvimento economico e o Nordeste Brasileiro*. ETENE publication no. 12. Fortaleza: Banco do Nordeste do Brasil, 1956.

*Legislação do DNOCS*. 2nd ed. Rio de Janeiro: Brasil-Ministerio de Viação e Obras Públicas, 1958.

Lima, José Otavio Pereira. *Terra Nordestina: Problemas, homens e fatos*. 2nd ed. Vol. 1 of *Coleção Mossoroense*. Mossoró, Rio Grande do Norte: Fundação Guimarães Duque, 1981. Originally published 1954.

Lisboa, Miguel Arrojado. "O problema das seccas." *Anais da biblioteca nacional do Rio de Janeiro* 35 (1913): 129–46.

Lobato, Monteiro. *Mr. Slang e o Brasil, e o problema vital*. São Paulo: Editora Brasiliense, 1964.

Luetzelburg, Philip von. *Estudo botânico do Nordeste*. Series 1A, publication no. 57. Rio de Janeiro: IFOCS, 1922.

Lutz, A., and A. Machado. "Viagem Pelo Rio São Francisco e por Alguns de Seus Afluentes . . . ." *Memórias do Instituto Oswaldo Cruz* 7 (1915): 5–62.

Lyra, A. Tavares de. *As secas do Nordeste*. Rio de Janeiro: Imprensa Nacional, 1919.

Meira, Lucio. *A luta contra as secas no Nordeste*. Rio de Janeiro: Minstério da Viação e Obras Públicas, 1958.

Mello, A. da Silva. *Nordeste Brasileiro: Estudo e impresses da viagem*. Rio de Janeiro: José Olympio, 1953.

Menezes, Djacir. *O outro Nordeste*. Rio de Janeiro: José Olympio, 1937.

Milanez, Alvaro. "Das atividades do Serviço Especial de Saúde Pública no campo do saneamento e da melhoria da habitação." *Revista do Serviço Especial de Saúde Pública* 10, no. 1 (1958): 325–57.

Motta, Lidenor de Mello. *Como resolver definitivamente o problema dos efeitos das secas no Nordeste*. Campo Grande, Mato Grosso do Sul: Livraria Ruy Barbosa, 1953.

Nash, Roy F. "Selling Public Health in Brazil: Five Years' Work of the International Health Board." *Brazilian American* 5, no. 123 (1922).

Neves, Antonio da Silva. *O maior problema econômico nacional: A sêcca de 1919: Uma questão velha por um prisma novo*. Rio de Janeiro: Officinas Graphicas do Jornal do Brasil, 1919.

Nobrega, Trajano Pires da. "Ensaio social-econômico de um setor do Vale do Rio São Francisco." *Boletim da Inspetoria Federal de Obras Contra as Sêcas* 16, no. 1 (1941): 3–14.

Parahym, Orlando. *O problema alimentar no sertão*. Recife: Imprensa Industrial, 1940.

Peixoto, Afranio. *Afranio vs. Afranio*. Nictheroy, Rio de Janeiro: Jeronymo Silva, 1922.

———. *Clima e doenças do Brasil*. Rio de Janeiro: Imprensa Nacional, 1907.

———. *Clima e saúde: Introdução bio-geográfica a civilização Brasileira*. Rio de Janeiro: Companhia Editora Nacional, 1938.

Penna, Belisário. *Defesa sanitária do Brasil*. Rio de Janeiro: Tipografia Revista dos Tribunais, 1922.

———. *O saneamento do Brasil: Sanear o Brasil é povoal-o; é enriquecel-o; é moralisal-o*. Rio de Janeiro: Typografia Revista dos Tribunães, 1918.

Penna, Belisário, and Arthur Neiva. "Viagem cientifica pelo norte da Bahia, sudoeste de Pernambuco, sul do Piauí e de norte a sul de Goiás." *Memórias do Instituto Oswaldo Cruz* 8, no. 3 (1916): 74–224.

Pereira, Miguel. "O Brasil é ainda um immenso hospital." *Revista de Medicina* 7, no. 21 (1922): 3–7.

Pessoa, Epitácio. *Obras completas*. 25 vols. Rio de Janeiro: Ministério da Educação e Cultura, Instituto Nacional do Livro, 1955.

———. *Obras do Nordeste: Resposta ao Exmo Snr. Senador Sampaio Corrêa*. Series C, vol. 587 of *Coleção Mossoroense*. Mossoró, Rio Grande do Norte: Fundacao Guimaraes Duque, 1990. Originally published 1925.

———. *Pela verdade*. Rio de Janeiro: Livraria Francisco Alves, 1925.

Peterson, John M. *Pesquisa econômica da TVA sobre desenvolvimento regional*. ETENE publication no. 14. Fortaleza: Banco do Nordeste do Brasil, 1956.

Raposo, Eduardo, and Humberto Melo. *Têcnicos do DNOCS:. Depoimentos de Genésio Martins, Antônio Gouveia Neto, João Alberto Gurgel, Paulo de Brito Guerra*. Rio de Janeiro: FGV/CPDOC História Oral, 1981.

Raposo, Eduardo, Humberto Melo, and Maria Antonia Alonso Andrade. *Firmino Leite depoimento, 1979*. Rio de Janeiro: FGV/CPDOC História Oral, 1981.

Reis, Aarão. *Relatório de obras novas contra as secas*. Rio de Janeiro, Imprensa Nacional, 1920.

Ribeiro, Raymundo F. "Valle do Rio Salgado e sua irrigação." *Revista do Instituto do Ceará* 35 (1921): 146–51.

Robock, Stefan H. *Brazil's Developing Northeast: A Study of Regional Planning and Foreign Aid.* Washington, DC: Brookings Institution, 1963.

——. "The Negro in the Industrial Development of the South." *Phylon: The Atlanta University Review of Race and Culture* 3 (1953): 319–25.

——. "What the United States Can Learn." In *Comparisons in Resource Management: Six Notable Programs in Other Countries and Their Possible U.S. Application,* edited by H. Jarrett, 251–59. Baltimore: Johns Hopkins University Press, 1961.

Rodrigues, J. A. Fonseca. "As seccas do Ceará: Ensaio theórico." *Boletim do Instituto Paulista de Engenharia* (1919).

Rodrigues, Raimundo Nina. *As collectividades anormaes.* Rio de Janeiro: Civilização Brasileira, 1939. Originally published 1897.

Rondon, C., P. de M. Barros, and I. S. Lopes. "Relatório apresentado ao Governo Federal pela Comissão Incumbida de Visitar as Obras Contra as Secas." *Revista Brasileira de Engenharia* 6, no. 2 (1923): 48–63.

Roquette-Pinto, E. *Grêmio Euclides da Cunha, por protesto e adoração: in memoriam.* Rio de Janeiro: Grêmio Euclides da Cunha, 1919.

Sarmiento, Domingo F. *Facundo, or, Civilization and Barbarism.* Translated by Mary Mann. New York: Penguin, 1998.

Serebrenick, Salomão. "A Comissão do Vale do São Francisco: Objetivos e realizações." *Revista Brasileira de Geografia* 22 (1960): 259–77.

Silva, Raymundo Pereira da. "O problema do norte." *Revista do Clube de Engenharia* 19 (1909): 9–109.

Silva, Rui de Lima e. "Geologia do estado de Pernambuco." *Revista do Instituto Archeologico e Geográfico Pernambucano* 33, nos. 155–58 (1933–35), 7–15.

Small, Horatio L. *Geologia e supprimento d'agua subterrânea no Ceará e parte do Piauhy.* Series 1D, publication no. 25. Rio de Janeiro: IOCS, 1913.

Smillie, Wilson G. "The Results of Hookworm Disease Prophylaxis in Brazil." *American Journal of Hygiene* 2 (1922): 77–95.

Sobrinho, Thomaz P. de Souza Brasil. "Açude 'Quixeramobim': Memória justificativa apresentada ao Exmo. Sr. Inspetor das Obras Contra as Secas." *Revista do Instituto do Ceará* 26 (1912): 215–79.

——. "Capacidade irrigatória do Açude Orós." *Revista do Instituto do Ceará* 41, nos. 1–4 (1927): 159–66.

——. "O homem do Nordeste." *Boletim da Inspetoria Federal de Obras Contra as Secas* 1, no. 6 (1934): 239–56.

——. "Povoamento do Nordeste Brasileiro." *Revista do Instituto do Ceará* 51 (1937): 107–62.

——. *O problema das seccas no Ceará.* 2nd ed. Fortaleza: Eugenio Gadelha, 1920.

——. "Terra das Secas." *Boletim da Inspetoria Federal de Obras Contra as Secas* 1, no. 5 (1934): 210–19.

Sopper, Ralph H. *Geologia e supprimento d'agua subterrânea em Sergipe e no Nordeste da Bahia.* Series 1G, publication no. 34. Rio de Janeiro: IOCS, 1914.

Souza, Eloy de. "Um problema nacional: projecto e justificação." In *Memorial da seca*, edited by Vingt-Un Rosado, 24–35. Vol. 158 of *Coleção Mossoroense*. Mossoró, Rio Grande do Norte: Fundaçao Guimaraes Duque, 1981. Originally published 1911.

Studart, Barão de. *Climatologia, epidemias e endemias do Ceará*. Fortaleza: Fundação Waldemar Alcântara, 1997.

Studart, Guilherme. *Diccionário bio-bibliográphico Cearense*. Vol. 3. Fortaleza: Typo-Lithographia a Vàpor, 1915.

Superintendência de Desenvolvimento do Nordeste, *A Policy for the Economic Development of the Northeast and a Synthesis of the First Guiding Plan for the Development of the Northeast*. Recife: SUDENE, 1961.

Teófilo, Rodolfo. *A fome: Scenas da sêcca do Ceará*. Fortaleza: Editora Gualter R. Silva, 1890.

——. *A Seca de 1915*. Rio de Janeiro: Imprensa Inglesa, 1922.

——. *Libertação do Ceará*. Lisbon: A Editora Limitada, 1914.

——. *Lyra rustica: Scenas da vida sertaneja*. Lisbon: Typographia da Editora Limitada, 1913.

——. *Seccas do Ceará: Segunda metade do seculo XIX*. Ceará, 1901.

——. *Varíola e vacinação no Ceará*. Fortaleza: Fundação Waldemar Alcântara, 1997.

Tigre, Carlos Bastos. *Catecismo do agricultor irrigante*. Fortaleza: DNOCS, 1954.

Torres, Alberto. *A organização nacional*. Rio de Janeiro: Impresa Nacional, 1914.

——. *O problema nacional Brasileiro*. Rio de Janeiro: Imprensa Nacional, 1914.

Trinidade, José Augusto. "Os postos agrícolas da inspetoria de sêcas." *Boletim da Inspetoria Federal de Obras Contra as Secas* 13, no. 2 (1940): 95–113.

——. *Os postos agrícolas da inspetoria de sêcas*. Rio de Janeiro: Ministério de Viação e Obras Públicas, 1940.

——. "Os serviços agricolas da inspectoria de seccas." *Boletim da Inspetoria Federal de Obras Contra as Secas* 7, no. 1 (1937): 27–43.

Vargas, Getúlio. *Presidential Message 363 of October 23, 1951*. Banco do Nordeste do Brasil publication no. 7. Fortaleza: Banco do Nordeste do Brasil, 1958.

Veloso, Ursulino Dantas. *O Nordeste seco e as obras contra as secas, 1910–1981*. Recife: Imprensa Universitaria/UFRPE, 1985.

Vieira, Luiz Augusto da Silva. "As Dificuldades com que lutou a Inspetoria de Sêcas, no socôrro aos flagelados." *Boletim da Inspetoria Federal de Obras Contra as Sêcas* 1, no. 4 (1934): 176–79.

——. "O ministro José Américo e o Nordeste." *A União* (João Pessoa, Pernambuco), August 3, 1934, 1.

——. "As obras contra as seccas no período dictatorial." *Revista do Clube de Engenharia* no. 5–10 (1935): 356–58.

——. "Obras do Nordeste." *Boletim da Inspetoria Federal de Obras Contra as Secas* 11, no. 1 (1939): 101–8.

——. "Obras no Nordeste." *Boletim da Inspetoria Federal de Obras Contra as Secas* 13, no. 2 (1940): 85–94.

Walker, Gilbert T. "Ceará Brazil Famines and the General Air Movement." *Beitrage zur Phys. der Freien Atmosphere* 14 (1928): 88–93.

Waring, Geraldo A. *Irrigation in Northeastern Brazil*. San Francisco: Western Engineering, 1912.

———. *Supprimento d'agua no Nordeste do Brasil*. Series 1D, publication no. 23. Rio de Janeiro: IOCS, 1923.


Secondary Sources

Abreu, Capistrano de. *Chapters of Brazil's Colonial History 1500–1800*. Translated by Arthur Brakel. New York: Oxford University Press, 1997.

Adas, Michael. *Dominance by Design: Technological Imperatives and America's Civilizing Mission*. Cambridge, MA: Belknap Press, 2006.

Albuquerque, Durval Muniz de, Jr. "Falas de astúcia e de angústia: A seca no imaginario nordestino, de problema a solução 1877–1922." Master's diss., Universidade Estadual de Campinas, 1987.

———. *A invenção do Nordeste e outras artes*. Recife: FJN Editora Massangana, 1999.

———. "Palavras que calcinam, Palavras que dominam: A invenção da seca do Nordeste." *Revista Brasileira de História* 15, no. 28 (1994–95): 111–20.

Alexander, Robert J. *Juscelino Kubitschek and the Development of Brazil*. Athens: Ohio University Center for International Studies, 1991.

Amado, Jorge. "Região, sertão, nação." *Estudos Históricos* 15 (1995): 145–51.

Amory, Frederic. "Euclides da Cunha and Brazilian Positivism." *Luso-Brazilian Review* 36, no. 1 (1997): 87–94.

Andrade, Francisco Alves de. *Ensino e desenvolvimento das ciências agrárias no Nordeste Ceará, 1918–1978*. Fortaleza: Banco do Nordeste do Brasil, 1979.

Andrade, Joaquim Marçal Ferreira de, and Rosângela Logatto. "Imagens da seca de 1877–78 no Ceará: uma contribuição para o conhecimento das origens do fotojornalismo na imprensa Brasileira." *Anais da Biblioteca Nacional* 114 (1994): 71–83.

Andrade, Manuel Correia de. *Geografia, sociedade e cultura*. Vol. 293 of *Coleção Mossoroense*. Mossoró, Rio Grande do Norte: Fundação Guimarães Duque, 1983.

———. *The Land and People of Northeast Brazil*. Translated by Dennis V. Johnson. Albuquerque: University of New Mexico Press, 1980.

Andrews, G. R. *Blacks and Whites in Sao Paulo, Brazil, 1888–1988*. Madison: University of Wisconsin Press, 1991.

Arnold, David. *Colonizing the Body: State Medicine and Epidemic Disease in Nineteenth-Century India*. Berkeley: University of California Press 1993.

Arons, Nicholas G. *Waiting for Rain: The Politics and Poetry of Drought in Northeast Brazil*. Tucson: University of Arizona Press, 2004.

Barbosa, José Policarpo de Araujo. *História da saúde pública do Ceará: Da colônia a Vargas*. Fortaleza: Edições UFC, 1994.

Birn, Anne-Emanuelle. "Local Health and Foreign Wealth: The Rockefeller Foundation's Public Health Programs in Mexico, 1924–1951." PhD diss., Johns Hopkins University, 1993.

———. *Marriage of Convenience: Rockefeller International Health and Revolutionary Mexico.* Rochester, NY: University of Rochester Press, 2006.

———. "A Revolution in Rural Health? The Struggle over Local Health Units in Mexico, 1928–1940." *Journal of the History of Medicine and Allied Sciences* 53 (1998): 43–76.

Birn, Anne-Emanuelle, and Armando Solórzano. "The Hook of Hookworm: Public Health and the Politics of Eradication in Mexico." In *Western Medicine as Contested Knowledge*, edited by Andrew Cunningham and Bridie Andrews, 147–71. Manchester, England: Manchester University Press, 1997.

Blake, Stanley. "The Invention of the Nordestino: Race, Region, and Identity in Northeastern Brazil, 1889–1945." PhD diss., State University of New York–Stony Brook, 2001.

———. *The Vigorous Core of Our Nationality: Race and Regional Identity in Northeastern Brazil.* Pittsburgh: University of Pittsburgh Press, 2011.

Bond, Kathleen. "A Drought Ravages Northeast Brazil." *NACLA Report on the Americas* 32 (1999): 11–13.

Borges, Dain. "Salvador's 1890s: Paternalism and Its Discontents." *Luso-Brazilian Review* 30, no. 2 (1993): 47–57.

Botelho, Caio Lóssio. *Seca: Visão dinâmica, integrada e correlações.* Fortaleza: ABC, 2000.

Brown, Larissa V. "Urban Growth, Economic Expansion, and Deforestation in Late Colonial Rio de Janeiro." In *Changing Tropical Forests*, edited by Richard P. Tucker and Harold Steen, 165–75. Durham, NC: Duke University Press, 1992.

Bursztyn, Marcel. *O poder dos donos: planejamento e clientelismo no Nordeste.* Petrópolis, Brazil: Vozes, 1984.

Campos, Andro Luiz Vieira de. "International Health Practices in Brazil: The Serviço Especial de Saúde Pública, 1942–1960." PhD diss., University of Texas at Austin, 1997.

Campos, Renato Carneiro. *Ideologia dos poetas populares.* 2nd ed. Recife: MEC, Instituto Joaquim Nabuco de Pesquisas Sociais, 1977.

Carey, Mark. "Climate and History: A Critical Review of Historical Climatology and Climate Change Historiography." *Wiley Interdisciplinary Reviews: Climate Change* 3 (2012): 233–49.

———. *In the Shadow of Melting Glaciers: Climate Change and Andean Society.* Oxford: Oxford University Press, 2010.

Carter, Eric D. *Enemy in the Blood: Malaria, Environment, and Development in Argentina.* Tuscaloosa: University of Alabama Press, 2012.

Carvalho, Otamar de. *A economia política do Nordeste: Secas, irrigação e desenvolvimento.* Brasília: ABID, 1988.

Castro, Lara de. *Avalanches de flagelados do sertão Cearense*. Fortaleza: DNOCS, BNB-ETENE, 2010.

Castro Santos, Luiz Antonio de. "Power, Ideology, and Public Health in Brazil, 1889–1930." PhD diss., Harvard University, 1987.

Cava, Ralph Della. *Miracle at Joaseiro*. New York: Columbia University Press, 1970.

Centro de Pesquisa e Documentação de História Contemporânea do Brasil, Fundação Getúlio Vargas, *Dicionário histórico-biográfico Brasileiro, 1930–1983*. 4 vols. Rio de Janeiro-RJ: Forense-Universitária, FINEP, 1984.

Chandler, Billy Jaynes. *The Feitosas and the Sertão dos Inhamuns: The History of a Family and a Community in Northeast Brazil, 1700–1930*. Gainesville: University Press of Florida, 1972.

Clarke, Kim A. *The Redemptive Work: Railway and Nation in Ecuador, 1895–1930*. Wilmington, DE: Scholarly Resource, 1998.

Cooper, Frederick, and Randall M. Packard, eds. *International Development and the Social Sciences: Essays on the History and Politics of Knowledge*. Berkeley: University of California Press, 1997.

Costa, Emilia Viotti da. *The Brazilian Empire: Myths and Histories*. Rev. ed. Chapel Hill: University of North Carolina Press, 2000.

Coutinho, Marilia. "Tropical Medicine in Brazil: The Case of Chagas' Disease." In *Disease in the History of Modern Latin America: From Malaria to AIDS*, edited by Diego Armus, 76–100. Durham, NC: Duke University Press, 2003.

Cueto, Marcos, ed. *Missionaries of Science: The Rockefeller Foundation and Latin America*. Bloomington: Indiana University Press, 1994.

Cukierman, Henrique Luiz. "Manguinhos, outras histórias: A tecnociência em terras Brasileiras." PhD thesis, Universidade Federal do Rio de Janeiro, 2001.

Cunniff, Roger. "The Great Drought: Northeast Brazil, 1877–1880." PhD diss., University of Texas at Austin, 1970.

Curran, Mark J. *A presença de Rodolfo Coelho Cavalcante na moderna literatura de cordel*. Rio de Janeiro: Editora Nova Fronteira, Fundação Casa de Rui Barbosa, 1987.

———. "Brazil's *Literatura de Cordel*: Poetic Chronicle and Popular History." *Studies in Latin American Popular Culture* 15 (1996): 219–29.

———. "Politics in the Brazilian *Literatura de Cordel*: The View of Rodolpho Coelho Cavalcante." *Studies in Latin American Popular Culture* 3 (1984): 115–26.

Dávila, Jerry. *Diploma of Whiteness: Race and Social Policy in Brazil, 1917–1945*. Durham, NC: Duke University Press, 2003.

Davis, Mike. *Late Victorian Holocaust: El Niño Famines and the Making of the Third World*. New York: Verso, 2001.

Dean, Warren. *Brazil and the Struggle for Rubber*. Cambridge: Cambridge University Press, 1987.

———. *The Industrialization of São Paulo, 1880–1945*. Austin: University of Texas Press, 1969.

———. *Rio Claro: A Brazilian Plantation System 1820–1920*. Stanford, CA: Stanford University Press, 1976.

———. *With Broadax and Firebrand: The Destruction of the Brazilian Atlantic Forest*. Berkeley: University of California Press, 1995.

Dia, G.M., N. Flowers, D. Gross, A. Mascarenhas, O. Mascarenhas, and E. Núnes. "Regional and Subregional Disjunctions: The Case of Northeastern Brazil." In Rolando V. Garcia and Pierre Spitz, eds., *Drought and Man*. Vol. 3, *The Roots of Catastrophe*, 75–117. Oxford: Oxford University Press, 1971.

Diacon, Todd A. *Stringing Together a Nation: Cândido Mariano da Silva Rondon and the Construction of a Modern Brazil, 1906–1930*. Durham, NC: Duke University Press, 2004.

*Dicionário histórico-biográfico Brasileiro, 1930–1983*. 4 vols. Rio de Janeiro: Forense-Universitária: FINEP, 1984.

Dinius, Oliver. *Brazil's Steel City: Developmentalism, Strategic Power, and Industrial Relations in Volta Redonda, 1941–1964*. Stanford, CA: Stanford University Press, 2011.

Diniz, Alberto. *O dinamismo patrioticamente construtivo de Belisário Penna: Esboço biográfico*. Rio de Janeiro: Jornal do Commércio, 1948.

Domingos, Manuel, and Laurence Hallewell. "The Powerful in the Outback of the Brazilian Northeast." *Latin American Perspectives* 31, no. 2 (2004): 94–111.

Duarte, Renato. *Do desastre natural à calamidade pública*. Recife: Fundação Joaquim Nabuco, 2002.

———. "A seca no Nordeste: de desastre natural a calamidade pública." *Trabalho para discussão*. Recife: Fundação Joaquim Nabuco, Instituto de Pesquisas Sociais, 1999.

Endfield, Georgina. *Climate and Society in Colonial Mexico: A Study in Vulnerability*. Malden, MA: Blackwell, 2008.

Ervin, Michael. "The 1930 Agrarian Census in Mexico: Agronomists, Middle Politics, and the Negotiation of Data Collection." *Hispanic American Historical Review* 87, no. 3 (2007): 537–70.

Escobar, Arturo. *Encountering Development: The Making and Unmaking of the Third World*. Princeton, NJ: Princeton University Press, 1995.

Faoro, Raymundo. *Os donos do poder: Formação do patronato político Brasileiro*,3rd ed. Vol. 2. Porto Alegre, Brazil: Editora Globo, 1976.

Farias, Lima Rodrigues de. "A fase pioneiro de reforma sanitária no Brasil: A atuação da Fundação Rockefeller, 1915–1930." Master's diss., Universidade Estadual do Rio de Janeiro, Instituto de Medicina Social, 1994.

Fausto, Boris. *A Concise History of Brazil*. New York: Cambridge University Press, 1999.

Ferguson, James. *The Anti-Politics Machine: "Development," Depoliticization, and Bureaucratic Power in Lesotho*. New York: Cambridge University Press, 1990.

Ferreira, Luis Otavio. "Os politécnicos: Ciência e reorganização social segundo o pensamento positivista da Escola Politécnica do Rio de Janeiro 1862–1922." Master's diss., Universidade Federal do Rio de Janeiro, 1989.

Fiege, Mark. *Irrigated Eden: The Making of an Agricultural Landscape in the American West.* Seattle: University of Washington Press, 1999.

Filho, Jorge Coelho da Silva. *As secas do Nordeste e a indústria das secas.* Petrópolis, Brazil: Editora Vozes, 1985.

Finan, T. J. "Climate Science and the Policy of Drought Mitigation in Ceará, Northeast Brazil." In *Weather, Climate, Culture,* edited by Sarah Strauss and Ben Orlove, 203–16. Oxford: Berg, 2003.

Fitzgerald, Deborah. *Every Farm a Factory: The Industrial Ideal in American Agriculture.* New Haven, CT: Yale University Press, 2003.

Froehlich, José Marcos. "Gilberto Freyre, a história ambiental e a 'rurbanização.'" *História, Ciências, Saúde—Manguinhos* 7 (2000): 283–303.

Furtado, Celso. *O Nordeste: Reflexões sobre uma política alternativa de desnvolvimento.* Coleção Mossoroense, series B, no. 649. Mossoró, Rio Grande do Norte: Coleção Mossoroense, 1989.

———. *Seca e Poder: Entrevista com Celso Furtado.* 2nd ed. São Paulo: Editora Fundação Perseu Abramo, 1999.

Gadelha, P. "Conforming Strategies of Public Health Campaigns to Disease Specificity and National Contexts: Rockefeller Foundation's Early Campaigns against Hookworm and Malaria in Brazil." *Parassitologia* 40 (1998): 159–75.

Garfield, Seth. *In Search of the Amazon: Brazil, the United States, and the Nature of a Region.* Durham, NC: Duke University Press, 2013.

Graham, Richard. *Britain and the Onset of Modernization in Brazil, 1850–1914.* Cambridge: Cambridge University Press, 1972.

———. *Patronage and Politics in Nineteenth-Century Brazil.* Stanford, CA: Stanford University Press, 1990.

Grant, Nancy L. *TVA and Black Americans: Planning for the Status Quo.* Philadelphia: Temple University Press, 1990.

Greenfield, Gerald Michael. *The Realities of Images: Imperial Brazil and the Great Drought. Transactions* 91 (2001). Philadelphia: American Philosophical Society.

———. "*Sertão* and *Sertanejo*: An Interpretative Context for Canudos." *Luso-Brazilian Review* 30, no. 2 (1993): 35–46.

Guedes, Zeito. "O folclore da seca." *Folclore* 216 (1991): 13–14.

Hall, Anthony M. *Drought and Irrigation in Northeast Brazil.* Cambridge: Cambridge University Press, 1978.

Hardiman, D. "The Politics of Water in Colonial India." *South Asia,* n.s., 25, no. 2 (2002): 111–20.

Hargreaves, M. W. *Dry Farming in the Northern Great Plains, 1900–1925.* Cambridge, MA: Harvard University Press, 1957.

———. *Dry Farming in the Northern Great Plains: Years of Readjustment, 1920–1990.* Lawrence: University of Kansas Press, 1993.

Hirschman, Albert O. *Journeys toward Progress: Studies of Economic Policy-Making in Latin America.* New York: Twentieth Century Fund, 1963.

Hochman, Gilberto. *A era do saneamento: As bases da politica de saúde pública no Brasil*. São Paulo: Hucitec/Ampocs, 1998.

——. "Logo ali, no final da avenida: *Os sertões* redefinidos pelo movimento sanitarista da Primeira República," supplement, *História, Ciência, Saúde— Manguinhos* 5 (1998): 218–35.

Katzman, Martin T. "The Brazilian Frontier in Comparative Perspective." *Comparative Studies in Society and History* 17 (1975): 266–85.

Kawamura, Lili Katsuco. *Engenheiro: Trabalho e ideologia*. 2nd ed. São Paulo: Editora Ática, 1981.

Kenny, Mary Lorena. "Landscapes of Memory: Concentration Camps and Drought in Northeastern Brazil." *Latin American Perspectives* 36, no. 5 (2009): 21–38.

Kirkendall, Andrew J. *Class Mates: Male Student Culture and the Making of a Political Class in Nineteenth-Century Brazil*. Lincoln: University of Nebraska Press, 2002.

——. *Paulo Freire and the Cold War Politics of Literacy*. Chapel Hill: University of North Carolina Press, 2010.

Kirkendall, Richard S. *Social Scientists and Farm Politics in the Age of Roosevelt*. Columbia: University of Missouri Press, 1966.

Klein, Marcus. *Our Brazil Will Awake! The Ação Integralista Brasileira and the Failed Quest for a Fascist Order in the 1930s*. Amsterdam: Cuadernos del Cedla, 2004.

Kropf, Simone Petraglia. *Doença de Chagas, doença do Brasil: Ciência, saúde e nação 1909–1962*. Rio de Janeiro: Editora Fiocruz, 2009.

——. "Sonho da razão, alegoria da ordem: O discuro dos engenheiros sobre a cidade do Rio de Janeiro no final do século XIX e início do século XX." In *Missionários do progresso: Médicos, engenheiros e educadores no Rio de Janeiro, 1870–1937*, edited by Micael Herschmann, Clarice Nunes and Simone Kropf, 69–154. Rio de Janeiro: Diadorim, 1996.

Leal, Victor Nunes. *Coronelismo: The Municipality and Representative Government in Brazil*. Translated by June Henfrey. Cambridge: Cambridge University Press, 1977.

Leprun, Jean-Claude. "Comparative Ecology of Two Semi-Arid Regions, the Brazilian Sertão and the African Sahel." In *Soils and Sediments: Mineralogy and Geochemistry*, edited by Hélène Paquet and Norbert Clauer, 157–72. Berlin: Springer, 1997.

Levine, Robert M. *Father of the Poor? Getulio Vargas and His Era*. Cambridge: Cambridge University Press, 1998.

——. *Vale of Tears: Revisiting the Canudos Massacre in Northeastern Brazil, 1893–1897*. Berkeley: University of California Press, 1992.

Levinson, Jerome. *The Alliance That Lost Its Way*. Chicago: Quadrangle, 1970.

Lewin, Linda. *Politics and Parentela in Paraíba: A Case Study of Family-Based Oligarchy in Brazil*. Princeton, NJ: Princeton University Press, 1987.

Li, Tania M. *The Will to Improve: Governmentality, Development and the Practice of Politics*. Durham, NC: Duke University Press, 2007.

Lima, Aline Silva. "Um projeto de 'combate às secas': Os engenheiros civis e as obras públicas na Inspetoria de Obras Contra as Secas (IOCS) e a construção do açude Tucunduba (1909–1919)." Master's diss., Universidade Federal do Ceará, 2010.

Lima, Nísia Trinidade. "Missões civilizatórias da República e interpretação do Brasil," supplement, *História, Ciências, Saúde—Manguinhos* 5 (1998): 163–93.

———. *Um sertão chamado Brasil: Intelectuais e representação geográfica da identidade nacional.* Rio de Janeiro: Revan/Iuperj, 1999.

Lima, N. T., and Gilberto Hochman. "Condenado pelo raça, absolvido pela medicina: O Brasil descoberto pelo movimento sanitarista da Primeira República." In *Raça, ciência e sociedade*, edited by Ricardo Santos and Marcos Chor Maio, 23–40. Rio de Janeiro: CCBB-Editora Fiocruz, 1996.

Lima, N. T., and N. Britto. "Salud y nación: Propuesta para el saneamento rural; Un estúdio de la revista *Saúde* 1918–1919." In *Salud, cultura y sociedad en América Latina: Nuevas perspectivas históricas*, edited by Marcos Cueto, 135–158. Lima: IEP-OPS, 1996.

———. *Saúde e nação: A proposta do saneamento rural; Um estudo da revista* Saúde. Vol. 3 of *Estudos de história e saúde*. Rio de Janeiro: Casa de Oswaldo Cruz, 1991.

Love, Joseph. "The Rise and Decline of Economic Structuralism in Latin America: New Dimensions." *Latin American Research Review* 40, no. 3 (2005): 100–125.

Löwy, Ilana. "Epidemiology, Immunology, and Yellow Fever: The Rockefeller Foundation in Brazil, 1923–1939." *Journal of the History of Biology* 30 (1997): 397–417.

———. "What/Who Should Be Controlled? Opposition to Yellow Fever Campaigns in Brazil, 1900–1939." In *Western Medicine as Contested Knowledge*, edited by Andrew Cunningham, 124–46. Manchester: Manchester University Press, 1997.

Madden, Lori. "The Canudos War in History." *Luso-Brazilian Review* 30, no. 2 (1993): 5–22.

Magalhães, A. R. "Drought and Policy Responses in Northeast Brazil." In *Drought Assessment, Management, and Planning: Theory and Case Studies*, edited by D. A. Wilhite 181–98. Boston: Kluwer Academic, 1993.

Magalhães, A. R., and M. H. Glantz, eds. *Socioeconomic Impacts of Climate Variations and Policy Responses in Brazil.* Brasilia: Esquel Brasil Foundation, 1992.

Magalhães, A. R., and P. Magee, "The Brazilian Northeast (Nordeste)." In *Drought Follows the Plow*, edited by Michael H. Glantz, 59–76. Cambridge: Cambridge University Press, 1994.

McCann, Bryan. *Hello, Hello Brazil: Popular Music in the Making of Modern Brazil.* Durham, NC: Duke University Press, 2004.

McCook, Stuart. "Promoting the 'Practical': Science and Agricultural Modernization in Puerto Rico and Colombia, 1920–1940." *Agricultural History* 751 (2001): 52–82.

———. *States of Nature: Science, Agriculture, and Environment in the Spanish Caribbean, 1760–1940*. Austin: University of Texas Press, 2002.

McLain, W. Douglas, Jr. "Alberto Torres, Ad Hoc Nationalist." *Luso-Brazilian Review* 4, no. 2 (1967): 17–34.

Meade, Teresa. *Civilizing Rio: Reform and Resistance in a Brazilian City, 1889–1930*. State College: Pennsylvania State University Press, 1997.

Medina, Eden. *Cybernetic Revolutionaries: Technology and Politics in Allende's Chile*. Cambridge, MA: MIT Press, 2011.

Melo, Marcus André B. C. de, et al. *Elites empresariais, processos de modernização e políticas públicas: O caso de Ceará*. Recife: Instituto de Estudos da Cidadania, 1994.

Metcalf, Alida. *Go-Betweens and the Colonization of Brazil, 1500–1600*. Austin: University of Texas Press, 2005.

Mitchell, Timothy. *Rule of Experts: Egypt, Techno-Politics, Modernity*. Berkeley: University of California Press, 2002.

Moraes, Kleiton de. *O sertão descoberto aos olhos do progresso: A Inspetoria de Obras contra as Secas 1909–1918*. Master's diss., Universidade Federal do Rio de Janeiro, 2010.

Mota, Leonardo. *Sertão alegre*. Fortaleza: Imprensa Universitária do Ceará, 1965.

Nachman, Robert G. "Positivism, Modernization and the Middle Class in Brazil." *Hispanic American Historical Review* 57, no. 1 (1977): 1–23.

Needell, Jeffrey D. "The Revolta Contra Vacina of 1904: The Revolt against 'Modernization' in Belle Epoque Rio de Janeiro." *Hispanic American Historical Review* 67, no. 2 (1987): 233–69.

Nelson, Donald R., and Timothy J. Finan. "Praying for Drought: Persistent Vulnerability and the Politics of Patronage in Ceará, Northeast Brazil." *American Anthropologist* 111, no. 3 (2009): 302–16.

Neto, Lira. *O poder e a peste: A vida de Rodolfo Teófilo*. Fortaleza: Edições Demócrito Rocha, 2001.

Neves, Frederico de Castro. "Getúlio e a seca: Políticas emergênciais na era Vargas." *Revista Brasileira de história* 21, no. 40 (2001): 107–31.

———."A seca na história do Ceará." In *Nova História do Ceará*, edited by Simone Souza et al., 76–102. Fortaleza: Editora Demócrito Rocha, 2000.

Nogueira, Marcos José et al. "Condições geo-ambientais do semi-árido Brasileiro." *Ciência e Trópico* 20, no. 1 (1992): 173–98.

Oliveira, Francisco de. *Elegia para uma re(li)gião: SUDENE, Nordeste; planejamento e conflitos de classes*. 2nd ed. Rio de Janeiro: Paz e Terra, 1977.

Oliveira, Lâucia Lippi. *A questão nacional na primeira república*. São Paulo: Editora Brasiliense, 1990.

Oliven, Ruben. *Tradition Matters: Modern Gaúcho Identity in Brazil*. New York: Columbia University Press, 1996.

Packard, Randall, and P. Gadelha. "A Land Filled with Mosquitoes: Fred L. Soper, the Rockefeller Foundation, and the Anophiles Gambiae Invasion of Brazil." *Medical Anthropology* 17 (1997): 215–38.

Page, Joseph A. *The Revolution That Never Was: Northeast Brazil, 1955–1964*. New York: Grossman, 1972.

Palmer, Steve. "Central American Encounters with Rockefeller Public Health, 1914–1921." In *Close Encounters of Empire: Writing the Cultural History of U.S.-Latin American Relations*, edited by Gilbert M. Joseph et al., 311–32. Durham, NC: Duke University Press, 1998.

Peard, Julyan G. *Race, Place, and Medicine: The Idea of the Tropics in Nineteenth-Century Brazilian Medicine*. Durham, NC: Duke University Press, 1999.

Pereira, Geraldo José Marques. *Aspectos economicos e sociais da saúde e da nutrição em Pernambuco*. Recife: UFPE, 1984.

Pessar, Patricia. *From Fanatics to Folk: Brazilian Millenarianism and Popular Culture*. Durham, NC: Duke University Press, 2004.

Phaelante, Renato. "A seca do Nordeste na poesia da música popular." *Ciência e Trópico* 21, no. 1 (1993): 97–121.

Phillips, Sarah T. *This Land, This Nation: Conservation, Rural America, and the New Deal*. Cambridge: Cambridge University Press, 2007.

Pisani, Donald. *Water and American Government: The Reclamation Bureau, National Water Policy, and the West, 1902–1935*. Berkeley: University of California Press, 2002.

Poppino, Rollie. "Cattle Industry in Colonial Brazil." *Mid-America* 31, no. 4 (1949): 219–47.

Rabelo, Fernanda Lima. *De experts a "bodes expiatórios": Identidade e formação da elite técnica do DASP e a reforma do serviço público federal no Estado Novo 1938–1945*. PhD thesis, Universidade Federal do Rio de Janeiro, 2013.

Reisner, Marc. *Cadillac Desert*. New York: Viking, 1986.

Ribeiro, Darcy. *Brazilian People: The Formation and Meaning of Brazil*. Translated by Gregory Rabassa. Gainesville: University Press of Florida, 2000.

Rios, Kênia Sousa. "O curral dos flagelados." *Revista Canudos, "Seca,"* 3, no. 1 (1999): 69–84.

Rodriguez, Julia. *Civilizing Argentina: Science, Medicine, and the Modern State*. Chapel Hill: University of North Carolina Press, 2006.

Roett, Riordan. *The Politics of Foreign Aid in the Brazilian Northeast*. Nashville: Vanderbilt University Press, 1972.

Rogers, Thomas. *The Deepest Wounds: A Labor and Environmental History of Sugar in Northeast Brazil*. Chapel Hill: University of North Carolina Press, 2010.

Rohter, Larry. "Brazilian Plan for Water Diversion Is Greeted by Skepticism." *New York Times*, March 28, 2005.

Sá, Lenilde Duarte de. "Parahyba: Uma cidade entre miasmas e micróbios. O serviço de higiene pública: 1895–1918." PhD diss., Escola de Enfermagem de Rebeirão Preto, 1999.

Safford, Frank R. *The Ideal of the Practical: Colombia's Struggle to Form a Technical Elite*. Austin: University of Texas Press, 1976.

Salgueiro, H. *Engenheiro Aarão Reis: O progresso como missão*. Belo Horizonte: Fundação João Pinheiro, 1997.

Santos, Martha. *Cleansing Honor with Blood*. Stanford, CA: Stanford University Press, 2012.

Schumacher, Ryan R. "The Rise and Decline of a Leather Civilization." Unpublished manuscript, 2003.

Schwartzman, Simon. *A Space for Science: The Development of the Scientific Community in Brazil*. University Park: Pennsylvania State University Press, 1991.

Scott, James C. *Seeing Like a State: How Certain Schemes to Improve the Human Condition Have Failed*. New Haven, CT: Yale University Press, 1998.

Sedrez, Lise. "Bay of All Beauties: State and Environment in Guanabara Bay, Rio de Janeiro, Brazil, 1875–1975." PhD diss., Stanford University, 2004.

Sen, Amartya. *Poverty and Famines: An Essay on Entitlement and Deprivation*. Oxford: Oxford University Press, 1981.

Sikkink, Kathryn. "Development Ideas in Latin America: Paradigm Shift and the Economic Commission for Latin America." In *International Development and the Social Sciences: Essays on the History and Politics of Knowledge*, edited by Frederick Cooper and Randall M. Packard, 228–58. Berkeley: University of California Press, 1997.

Silva, Roberto. "Entre dois paradigmas: Combate à seca e convivência com o semi-árido." *Sociedade e Estado* 18, nos. 1–2 (2003): 361–85.

Skidmore, Thomas E. *Black into White: Race and Nationality in Brazilian Thought*. Durham, NC: Duke University Press, 1993.

———. *Brazil: Five Centuries of Change*. 2nd ed. New York: Oxford University Press, 2010.

———. *Politics in Brazil, 1930–1964*. New York: Oxford University Press, 1967.

———. "Racial Ideas and Social Policy in Brazil, 1870–1940." In *The Idea of Race in Latin America, 1870–1940*, edited by Richard Graham, 7–36. Austin: University of Texas Press, 1990.

Slater, Candace. *Stories on a String: The Brazilian Literatura de Cordel*. Berkeley: University of California Press, 1982.

Soluri, John. *Banana Cultures: Agriculture, Consumption, and Environmental Change in Honduras and the United States*. Austin: University of Texas Press, 2005.

Sousa, João Morais de. "Discussão em torno do conceito de coronelismo." *Cadernos de Estudos Sociais* 11, no. 2 (1995): 321–35.

Souto, Francisco. *Nordeste: Poder subdesenvolvimento sustentado; Discurso e prática*. Fortaleza: Editora UFCE, 1992.

Souza, Candice V. E. *A pátria geográfica: Sertão e litoral no pensamento social Brasileiro*. Goiânia, Goiás: Editora da UFG, 1997.

Souza, Vanderlei Sebastião de. "Artur Neiva e a questão nacional nos anos 1910 e 1920." *História, Ciências, Saúde—Manguinhos* 16, S1 (2009): 249–64.

Stepan, Nancy Leys. *The Beginnings of Brazilian Science: Oswaldo Cruz, Medical Research and Policy, 1890–1920.* New York: Science History, 1976.

——. "Eugenics in Brazil." In *The Wellborn Science: Eugenics in Germany, France, Brazil, and Russia,* edited by Mark B. Adams, 110–52. Oxford: Oxford University Press, 1990.

——. *"The Hour of Eugenics": Race, Gender, and Nation in Latin America.* Ithaca, NY: Cornell University Press, 1991.

——. " 'The Only Serious Terror in These Regions': Malaria Control in the Brazilian Amazon." In *Disease in the History of Modern Latin America: From Malaria to AIDS,* edited by Diego Armus, 32–41. Durham, NC: Duke University Press, 2003.

——. *Picturing Tropical Nature.* Ithaca, NY: Cornell University Press, 2001.

——. "Portraits of a Possible Nation: Photographing Medicine in Brazil." *Bulletin of the History of Medicine* 68 (1994): 136–49.

Stone, Ian. "Canal Irrigation and Agrarian Change: The Experience of the Ganges Canal Tract, Muzaffarnagar District U.P., 1840–1900." In *Agricultural Production and Indian History,* edited by David Ludden, 114–44. Delhi: Oxford University Press, 1994.

Swarnakar, Sudha. "Drought, Misery and Migration: The Fictional World of José Américo de Almeida's *A Bagaceira* and Jorge Amado's *Seara Vermelha* and *Gabriela, Cravo e Canela.*" Paper presented to the Latin American Studies Association, Dallas, March 27–29, 2003.

Teixeira, L. A. "Da raça à doença em *Casa-grande e senzala.*" *História, Ciências, Saúde—Manguinhos* 4, no. 2 (1997): 231–43.

Telles, Edward. *Race in Another America: The Significance of Skin Color in Brazil.* Princeton, NJ: Princeton University Press, 2004.

Tendler, Judith. *Good Government in the Tropics.* Baltimore: Johns Hopkins University Press, 1997.

Thielen, Eduardo Vilela, Fernando Antonio Pires Alves, Jaime Larry Benchimol, Marli Brito de Albuquerque, Ricardo Augusto dos Santos, and Wanda Latmann Weltman, eds. *Science Heading for the Backwoods: Images of the Expeditions Conducted by the Oswaldo Cruz Institute Scientists to the Brazilian Hinterland, 1911–1913.* Rio de Janeiro: Instituto Oswaldo Cruz, 1991.

Tobelem, Alain. *Josué de Castro e a descoberta da fome.* Rio de Janeiro: Editora Leitura, 1974.

Turazzi, Maria Inez. *A eufória do progresso e a imposição da ordem: A engenharia, a indústria, e a organização do trabalho na virada do século XIX ao XX.* Rio de Janeiro: COPPE, 1989.

Vargas Llosa, Mario. *The War of the End of the World.* New York: Farrar, Straus and Giroux, 1984.

Vaughan, Meghan. *Curing Their Ills: Colonial Power and African Illness.* Oxford: Polity, 1991.

Vieira, Sulamita. *O sertão em movimento: A dinâmica da produção cultural.* São Paulo: Annablume, 2000.

Villa, Marco Antonio. *Vida e morte no sertão: História das secas no Nordeste nos séculos XIX e XX*. São Paulo: Editora Ática, 2000.

Wagner, Robert. *A conquesta do oeste: A fronteira na obra de Sérgio Buarque de Holanda*. Belo Horizonte: Editora UFMG, 2000.

Wahrlich, Beatriz M. de Souza. *Reforma administrativa na era de Vargas*. Rio de Janeiro: Editora de Fundação Getúlio Vargas, 1983.

Webb, Kempton. *The Changing Face of Northeast Brazil*. New York: Columbia University Press, 1974.

Weinstein, Barbara. *For Social Peace in Brazil: Industrialists and the Remaking of the Working Class in São Paulo, 1920–1964*. Chapel Hill: University of North Carolina Press, 1996.

———. "Racializing Regional Difference: São Paulo versus Brazil, 1932." In *Race and Nation in Modern Latin America*, edited by Nancy P. Appelbaum, Anne S. Macpherson, and Karin Alejandra Rosemblatt, 237–62. Chapel Hill: University of North Carolina Press, 2003.

Whitcombe, E. "The Environmental Costs of Irrigation in British India: Waterlogging, Salinity, Malaria." In *Nature, Culture, Imperialism: Essays on the Environmental History of Southeast Asia*, edited by David Arnold and Ramachandra Guha, 237–59. Delhi: Oxford University Press. 1995.

White, Richard. *"It's Your Misfortune and None of My Own": A History of the American West*. Norman: University of Oklahoma Press, 1991.

Williams, Daryle. *Culture Wars in Brazil: The First Vargas Regime, 1930–1945*. Durham, NC: Duke University Press, 2001.

Williams, Steven C. "Nationalism and Public Health." In *Missionaries of Science: The Rockefeller Foundation and Latin America*, edited by Marcos Cueto, 23–51. Bloomington: Indiana University Press, 1994.

Wolfe, Mikael. "Bringing the Revolution to the Dam Site: How Technology, Labor, and Nature Converged in the Microcosm of a Company Town in 1930s and 40s Mexico." *Journal of the Southwest* 53, no. 1 (2011): 1–31.

———. *Watering the Revolution: An Environmental and Technological History of Agrarian Reform in Mexico*. Durham, NC: Duke University Press, 1917.

Worster, Donald. *Dust Bowl: The Southern Plains in the 1930s*. New York: Oxford University Press, 1979.

———. *Rivers of Empire: Water, Aridity, and the Growth of the American West*. New York: Pantheon, 1985.

———. *Under Western Skies: Nature and History in the American West*. Oxford: Oxford University Press, 1992.

Wright, Angus, and Wendy Wolford. *To Inherit the Earth: The Landless Movement and the Struggle for a New Brazil*. Oakland, CA: Food First, 2003.

# Index

Note: Page numbers in italics refer to figures.

*A Bagaceira* (Almeida), 125–27, 134
Abolition, 20, 52, 57, 69
Academia Nacional de Medicina, 65, 66
Ação Integralista Brasileira (AIB; Brazilian Integralist Action), 130
Accioly, Antônio Pinto Nogeuira, 37, 54
Acculturation, 147, 154–55
Afro-Brazilians: prejudice toward, 77–79; as slaves, 199–200; whitening and, 57–58, 78. *See also* Blacks; Race
Agassiz, Louis, 32
Agency for International Development (U.S.), 211
Agrarian reform, 193–98
Agricultural posts, 149–50, *151*, 159–61, *160*, 161–62
Agriculture, 18–19, 25–26, 38–39, 152–59, 185–86. *See also* Food production; Irrigation
Agriculture training, 162–63
Agronomy, 3, 9–10, 132–33, 140–45, 147, 157–68, 222
Albano, Ildefonso, 103, 112, 114
Albuquerque, Durval Muniz de, Jr., 23
Albuquerque, João Pedro de, 56
Alencar, José Delfino, 166
Aliança Liberal (Liberal Alliance), 141
Aliança Nacional Libertadora (National Liberation Alliance), 130
Alliance for Progress, 207, 209–10
Almeida, José Américo, 125–27, 131, 134–42, 183, 239n21

Alves, Francisco de Paula Rodrigues, 37, 38, 54
Amado, Jorge, 68
Amaral Peixoto, Ernesto do, 196, 204
Andrade, Francisco Alves de, 196, 217
Araújo, Alarico, 82
Arraes, Miguel, 210
Atahyde, João Martins, 174

Banco do Nordeste do Brasil (BNB; Bank of Northeast Brazil), 183–87, 188, 201
Banco Nacional de Desenvolvimento Econômico (BNDE; National Economic Development Bank), 183, 188
Banditry, 109–10
Barbosa, J. A. de Castro, 100
Barreira, Inacio, 159
Barros, Gomes de, 43, 119
Barros, Paulo de Moraes, 117, 118, 235n55, 237n101
Beef, 18, 135, 151
Belém-Fortaleza Highway, 204
Beltrão, Oscar, 197
Bernardes, Artur, 116–17
Berredo, Vinicus, 181
Blacks: in Ceará, 19; immigration policy and, 78; as inferior, 126; International Health Board and, 47; sertão and, 22, 68; whiteness and, 129. *See also* Afro-Brazilians; Race
Blake, Stanley, 30, 129

Borges, Dain, 27
Bouchardet, Joanny, 100
Braga, Cincinato, 114–16
*Branqueamento*, 57–58, 59, 78, 128
Brás, Wenceslau, 66
Brasil, Thomas Pompeu de, 32–33
Brasileiro, Lloyd, 132
Brasil Sobrinho, Thomaz Pompeu de
  Souza, 96–97, 100, 112–14, 120–21,
  129, 154, 236n68
Brito Guerra, Paulo de, 144
Bureau of Reclamation (U.S.), 3, 9,
  37–38, 100–101, 104, 111–12, 155–56,
  235n48

*Caboclo*, 29, 128, 160
Callado, Antônio, 198–200
Câmara, Dom Helder, 188
Campina Grande, 50
*Campos de asistência*, 136
Canudos, 26–32
Cárdenas, Lázaro, 164–65
Carey, Mark, 6
Carnauba palm wax, 19
Carneiro, B. Piquet, 87–88
Castanhão Dam, 219
Castelo Branco, Humberto de Alencar,
  211
Castro, Fidel, 207, 209
Castro, Josué de, 170–71, 188–89, 191
*Catecismo do Agricultor* (Tigre), 163
Catholic Church, 62, 130, 136, 188,
  210–11
Cattle, 17–19, 91. *See also* Beef
Cavalcante, Rodolpho Coelho, 168,
  170
Cavalcanti, José, 49
Cedro Reservoir, 92–94, 102
Chagas, Carlos, 54, 72, 76, 101
Chagas disease, 67–68
Cholera, 24, 51–52
Choró Reservoir, 165–66
Citizenship, 54–60, 55

Civil engineering, 3, 9–10; agronomy
  vs., 140–45; development and, 85–88;
  frustrations with, 83–84; history of,
  86; middle politics and, 105–10;
  migrants in, 82–83; pay in, 107–8;
  Pessoa and, 110–17; positivism and,
  87. *See also* Irrigation
Civilization: in da Cunha, 28; environment and, 29–30; sanitation and, 68;
  sertão and, 61; whiteness and, 23
Climate, 89–91; animal husbandry
  and, 151; in da Cunha, 28; disease
  and, 56–57; drought and, 43;
  engineering and, 222; in sertão, 16;
  vulnerability and, 123, 218. *See also*
  Drought
Clube de Economistas, 192
Clube de Engenharia, 86–87, 98–99,
  137
Coexistence, 219
Coffee, 57, 193–94
Cold War, 208–9
Colonialism, 56
Comissão de Serviços Complementares
  da Inspetoria da Secas (CSC;
  Commission of Services Complemen-
  tary to the Inspectorate for
  Droughts), 150–54
Comissão do Vale do São Francisco
  (CVSF; São Francisco Valley
  Commission), 181–82, 188
Comissão Técnica de Reflorestamento e
  Postos Agrícolas do Nordeste
  (Northeast Technical Commission
  for Reforestation and Agricultural
  Posts), 148–49
Companhia de Desenvolvimento do
  Vale do São Francisco, 214
Companhia Hidrelétrica de São
  Francisco (CHESF; São Francisco
  Hydroelectric Company), 182
Comte, Auguste, 87
Concentration camps, 19, 24

Conselheiro, Antônio, 26–28, 29, 228n22
Conselho de Desenvolvimento do Nordeste (CODENO; Economic Development Council for the Northeast), 193–201
*Convivência,* 219
*Cordéis,* 41–43, 168–71
Coronéis, 34–36, 42, 92, *161*
*Coronelismo,* 34–35
Costa, Emilia Viotti da, 88
Cotton, 19–20, 41, 50, 111
Coup, 11–12, 87, 210–11, 214
Crandall, Roderic, 95
Crime, 109–10
Cruz, Oswaldo, 11, 38, 54, 66, 68
Cueto, Marcos, 6
"Culture managers," 128
Cunniff, Roger, 32
Curran, Mark, 42
C. W. Walker & Co., 111

da Cunha, Euclides, 27–30, 31–32, 37, 43, 125, 238n2
Dantas Veloso, Ursulino, 121
Dávila, Jerry, 129
DDT, 190
Dean, Warren, 6
Departamento de Administração do Serviço Público (DASP; Department of Administration of Public Services), 143–44
Departamento Nacional de Endemias Rurais (National Department of Rural Diseases), 187
Departamento Nacional de Obras Contra as Secas (DNOCS; National Department for Works to Combat Droughts), 1–3, 9, 38, 40, 144–48, 223; Banco do Nordeste do Brasil and, 184; Callado and, 198–99; CODENO and, 196–97; criticism of, 182–83; decline in credibility of,

180–83; Kubitschek and, 187–88; Orós Dam and, 204–5; reservoirs and, 39, 163–64; SUDENE and, 200–201; technocracy and, 176–77; Tigre and, 163
Departamento Nacional de Saúde Pública (DNSP) (National Department of Public Health), 72–74, 187
Derby, Orville, 95
Development: as acculturation, 147, 154–55; botanical species and, 62–63; civil engineering and, 85–88; context of, 15–16; criticism of, 41–43; disillusion and, 145–46; drought aid as failure in, 37–41; economic, 183–87, 191–98; federal investment in, 97–105; foreign models of, 179–80; Kubitschek and, 187–90; limitations of, 176–77; medicine and, 38; middle politics and, 105–10; patronage and, 35–37; Pessoa and, 110–17; reservoirs and, 91–92, 142–43
Diet, 151, 183, 188–89, 217. *See also* Malnutrition
Diretoria Geral de Saúde Pública (General Directorate for Public Health), 66
Discrimination, 77–79. *See also* Race
Disease, 47–49, 51–57, 65–66, 69, 76–77, 190–91. *See also* Public health
Doctors, 77; DDT and, 190; public health and, 76–77; race and, 61; vaccinations and, 53
Dominican friars, 62
Drought, 8–10; aid as development failure, 37–41; alleviation, 32–37, *33*; declaration of, 16; prediction, 29; "rainmaker" school and, 32, 90; severe, 16; as social phenomenon, 43–44; theories of, 89–91. *See also* Great Drought of 1877–79
Drought of 1958, 190–93
"Drought polygon," 180

Dry farming, 155–56
Duarte, Renato, 38
Duque, José Guimarães, *128*, 148, 153–58, 181, 194, 199, 202, 214
Dutra, Marechal Eurico Gaspar, 175, 181
Dwight P. Robinson & Co., 111

Economic Commission for Latin America (ECLA), 191–92
Economic development, 183–87, 191–98. *See also* Development
Education, 156–57, 186–87. *See also* Medical schools
Egypt, 99–100, 104
Electricity, 39–40, 111, 162, 181–82, 202, 211
Empresa Brasileira de Assistencia Tecnica e Extensao Rural (EM-BRATER, Brazilian Enterprise of Technical Assistance and Rural Extension), 215
Engineering. *See* Civil engineering
Epidemics. *See* Disease
Ervin, Michael, 6
Escobar, Arturo, 5
Escola de Minas, 86, 96
Escola Militar, 86
Escola Nacional de Agronomia, 149
Escola Politécnica, 86, 233n4
Escritório Técnico de Estudos Econômicos do Nordeste (ETENE; Technical Office for Economic Studies of the Northeast), 185–86, 191

Famine, 34, 100, 178–79, 188–89
Faoro, Raymundo, 35
Farias, Lima Rodrigues de, 76
Farming. *See* Agriculture; Agronomy; Food production; Irrigation
Feast of São José, 29
Ferguson, James, 5
Finan, Timothy, 218
Fish stocking, in reservoirs, 135

Flu, 65–66
*Folhetos*, 41–42
Folk music, 172–176
Fonseca Rodrigues, J. A., 102
Food and Agriculture Organization, 186
Food production, 50, 158, 188–89, 217–18. *See also* Agriculture; Irrigation
*Fornecedores*, 9
Forquilha Reservoir, 106–7, 147, 159, 167
Fournier, Luiz Mariano de, 90
Freire, Paulo, 210–11
Freyre, Gilberto, 32, 202
Furtado, Celso, 8–9, 12–13, 178–79, 188, 191–98, 200–208, 211–13, 217–18

Getúlio Vargas Foundation, 178, 191
Gold, 18
Gomes Coêlho, Silvestre, 167–68
Gonzaga, Luiz, 172–75
Goulart, João, 13, 187, 206, 207–9
Graham, Richard, 35–36
Great Drought of 1877–79, 10, 19, 24, 51–53, 85. *See also* Drought
Greenfield, Gerald, 30–31, 34
Grupo de Trabalho para o Desenvolvimento do Nordeste (GTDN; Working Group for Northeast Development), 188–90, 202

Hackett, Lewis W., 74
Hall, Anthony, 215–16
Haussmann, Georges-Eugène, 66
Health infrastructure, 48–54. *See also* Doctors; Medicine
Health surveys, 46, 54–56, *55*, 60–65, *64*, 124
Hirschmann, Albert, 37
Hochman, Gilberto, 67, 69
Hookworm, 67–68, 77, 232n81
Hydroelectricity, 182. *See also* Electricity

Immigration, 23, 49, 59, 67–68, 78–79, 96, 231nn33, 36
India, 99, 102, 104
Individualism, 154, 161
Industrialization, 31–32, 68
Inequality, 166–67. *See also* Race
Influenza, 65–66
Inspetoria de Obras Contra as Secas (IOCS; Inspectorate for Works to Combat Droughts), 11, 37, 56, 83, 84–85; Almeida and, 126–27, 135; authority over, 140–45; birth of, 82–124; Braga and, 114–16; budget of, 94–95, 138–39; Carneiro and, 87–88; Cedro Reservoir and, 92–94; criticism of, 41; development and, 85–88; DNOCS and, 144; Duque and, 155–56; establishment of, 85–86; farming regimes and, 150; funding of, 112; impact of, 109–10, 239n39; irrigation and, 88–97, 145–46, 149–50; Lisboa and, 89–97; Orós Dam and, 103, 120–21; and other problems in sertão, 123–24; Passoa and, 111; Reis and, 101–2; renaming of, 111; reservoirs and, 38, 95–96; Trinidade and, 148–50; Vargas and, 132, 133; Vieira and, 135–36; Waring and, 95–96
Inspetoria Federal de Obras Contra as Secas (IFOCS; Federal Inspectorate for Works to Combat Droughts). *See* Departamento Nacional de Obras Contra as Secas (DNOCS; National Department for Works to Combat Droughts); Inspetoria de Obras Contra as Secas (IOCS; Inspectorate for Works to Combat Droughts)
Institute of Inter-American Affairs, 182
Instituto Histórico-Geográfico Brasileiro, 31
Instituto Oswaldo Cruz, 11, 54–56

International Health Board (IHB), 46, 47–48, 74–80, 77
International Monetary Fund (IMF), 187, 207
Irrigation, 104–5, 111–18, 139–43; agricultural training and, 162–63; Alves and, 37; CODENO and, 194–95, 198–201; diet and, 152–53; in Egypt, 100; engineering and, 85; farming and, 149–50; federal authority and, 99; IFOCS and, 145–46, 149–50; inequality and, 166–67; IOCS and, 88–97; land redistribution and, 40; reservoirs and, 39, 142–43, 181; smallholder, 93, 145, 158–64; in Tunisia, 43; United States and, 100–101. *See also* Reservoirs

Julião, Francisco, 197–98, 200

Kennedy, John F., 207, 209
Kubitschek, Juscelino, 12, 163–64, 187–90, 198, 200, 205

Labor protections, 208
Lagoa do Peixe, 82
Lamarckianism, 29, 56, 58
Land Law of 1850, 19
Land rentals, 161–62
Leal, Victor Nunes, 35
Levine, Robert, 29
Lewin, Linda, 50, 109–10
Li, Tania, 6–7, 88
Liberal Alliance, 141
Liberal Party, 32–33
Liga Pro-Saneamento (Pro-Sanitation League), 67–74
Lima Campos, Artur Fragoso de, 135–36
Lima e Silva, Rui de, 133
Lisboa, Miguel, 89–97, 101, 117, 121, 122, 203
Literacy, 50–51, 186–87

Lopes, Ildephonso Simões, 117, 119, 140–41, 168, 237n101
Lula da Silva, Luiz Inácio, 219–20
Lutz, Adolpho, 56

Machado, Astrogildo, 56, 68
Malaria, 67–68
Malnutrition, 50, 63–64, 69, 108, 131
Manso, Cordeiro, 21–22
Maranhão, 76–77
Marginality, 21–23
Marxism, 198, 210–11
Masculinity, 26
McCook, Stuart, 6
Medical schools, 54, 56, 86
Medicine: development and, 38; embrace of modern, 56–57; paternalism and, 61; sanitation and, 71–72; science and, 66–67; SUDENE and, 208; "tropical," 56–57. *See also* Disease; Doctors; Public health; Vaccines
Medina, Eden, 6
Meira, Lucio, 190
Mello, Antônio da Silva, 183
*Mestiço*, 22, 30, 51, 58, 78, 118, 126. *See also* Afro-Brazilians; Race
Mexico, 6, 106, 154
"Middle politics," 4, 6, 105–10, 124, 147, 164–65
Millenarianism, 10
Ministério de Viação e Obras Públicas (Ministry of Transportation and Public Works), 125
Miscegenation, 78–79
Mitchell, Timothy, 5
Modernization, 21–23, 31–32, 79, 80, 127, 223–24
Mosquitoes, 63
Mota, Leonardo, 41
Music, 172–176
Mysticism, 28–29

National Agriculture Society, 141
Neiva, Arthur, *17*, 47, 54–56, *55*, 60–65, 124
Nelson, Donald, 218
New Deal, 178
Nile River, 99–100
Nobrega, Trajano Pires da, 159–62
Nordeste, 8, 11; emergence of, as distinct region, 15; Vargas in, 131–32
Norton Griffiths & Co., 111
Nutrition, 183, 188–89, 217, 236n68. *See also* Malnutrition

Obras Novas Contra as Secas (New Works to Combat Drought), 101–2
Orós Dam, 103, 120–21, 172, 187, 203–5, 219
*Os Sertões (Rebellion in the Backlands)* (da Cunha), 28, 29, 37

Palhano de Jesus, José, 122
Palmer Drought Severity Index, 16
Paraguayan War, 20
Paraíba, 50–51, *77, 160*
Parente Pessoa, José Cândido Castro, 196–97, 199, 204
Paris, 66
Partido Trabalhista Brasileiro (Brazilian Labor Party), 205, 206
Pasteur Institute, 73–74
Paternalism, 61, 70, 127–34, 148
Patronage, political, 32–37, *33*, 218
*Paulistas*, 23, 28, 57, 127–30
Paulo Afonso waterfall, 175, 182
Peasant Leagues, 197–98, 200, 208–9, 210, 216, 245n43
Peixoto, Afranio, 56
Penna, Belisário, 11, *17*, 221–22; Ação Integralista Brasileira and, 130; Departamento Nacional de Saúde Pública and, 72–73; health surveys and, 46, 54–56, *55*, *64*, 124; modernization and, 80; public health and,

47–48, 56–57, 67–72, 79–80; race and, 57–58; Reis and, 101; Rose vs., 79–80; and rural aid, 46–47; Rural Prophylaxis Service and, 72; and rural sanitation, 65–67

Pereira, Miguel, 71

Silva, Raymundo Pereira da, 98–99

Pernambuco, 48–49, 76–77

Pessoa, Epitácio, 71–72, 110–22, 123, 140–41, 203

Pessoa, João, 134

Pires, Acácio, 71

Pisani, Donald, 104

Plague, 50

Poetry, 41–43, 133, 168–70, 175–76

Point IV Mission (U.S.), 186

*Politécnicos*, 86–87

Political patronage, 32–37, *33*, 218

POLONORDESTE (O Programa de Desenvolvimento de Areas Integradas do Nordeste), 214

Positivism, 11, 26, 27, 87

Poverty, 31, 74, 105, 130, 172, 185–86, 219–21

Prebisch, Raul, 12, 179

Projeto Sertanejo, 214

Property holdings, 20–21

Pro-Sanitation League, 67–74

PROTERRA (Programa de Redistri- buição de Terra e de Estímulo à Agroindústria do Norte e do Nordeste) credit program, 214, 216

Public health, 46–48, 51–57, 65–80, 77–78, 182, 190–91. *See also* Doctors; Medicine

Quadros, Jânio, 205–6

Queiroz, Raquel de, 130

Quixadá Reservoir, 33, *33*, 122

Rabies, 73–74

Race: in Almeida, 126; in da Cunha, 28, 29, 31–32, 84, 238n2; doctors and, 61;

hierarchy of, 57–58; in media, 147; modernization and, 22–23, 58; Nobrega and, 160–61; in Penna, 47; in polygeny, 58; public health and, 57–58, 77–78; in Rose, 77–78; sertão and, 22–23, 68; Vargas and, 128–29. *See also* Afro-Brazilians; Blacks; *Mestiço*; Whiteness; Whitening

Railroad extensions, 33–34

Rainfall, 90–91

"Rainmaker" school, 32, 90

Ramos, Graciliano, 130–31

Ranches, 17–18, 38–39

Reforestation, 32

Refugee camps, 136, 182

Regionalism, 23

Reis, Aarão, 7, 101–2, 236n72

Reisner, Marc, 104

Reservoirs, 33, *64*, *84*; agronomy and, 166–68; Almeida and, 135–36, 183; Braga and, 114–16; Brasil Sobrinho and, 112–14; CODENO and, 194–95, 199; continued focus on, 219–20; da Cunha and, 43; development and, 91–92, 142–43; DNOCS and, 39, 163–64; and employment of refu- gees, 137, 236n78; engineering and, 85, 93–94, 127; food production and, 158; IFOCS and, 138–39; as insuffi- cient, 63; IOCS and, 38, 95–96; irrigation and, 39, 142–43, 181; under Kubitschek, 163–64; land expropria- tion and, 123, 155; Lisboa and, 91–93, 97, 111–12; "middle politics" and, 105–10; Pessoa and, 114–21; political patronage and, 32–33; "rainmaker thesis" and, 90; small, *64*, 91; social transformation and, 158–59; stocking of, with fish, 135; Trinidade and, 153; Vargas and, 132. *See also* Irrigation

Revy, Julian J., 33, 99–100

Ribeiro, Costa, 53

Ribeiro, Darcy, 40

Robinson, Dwight P., 103–4, 111
Robock, Stefan, 184, 185–87, 202, 206–7
Rocha, José Moreira da, 73
Rockefeller Foundation, 3, 11, 46, 74–80, 77, 221
Rodriguez, Julia, 6
Rondon, Cândido, 117, 118, 119, 237n101
Roosevelt, Franklin Delano, 178
Roquette-Pinto, Edgard, 31–32, 68
Rose, Wickliffe, 46, 47, 77–80
Rubber, 234n36
*Rule of Experts* (Mitchell), 5
Rural Prophylaxis Service, 66, 67, 72, 76

St. Lucia's Day, 29
Sales, Apolônio, 143–44, 182
*Saneamento do Brasil* (Penna), 68
Sanitation, 53–54, 65–67, 71–72
Santa Casa de Misericórdia, 25
Santo Antônio Reservoir, 82, *84*
Santos, Abelardo dos, 106–9
Santos, Luiz Antonio de Castro, 74
Santos, Martha, 19–21, 26
São Francisco River, 32, 233n17
São Francisco River system, 95
São Gonçalo agricultural post, *128*
São Paulo, 23
*Saúde* (journal), 68
Schools, 156–57
Schwartzman, Simon, 86
Science: health surveys and, 62–63; modernization and, 80; paternalism and, 70; social history of, 5–6; technocracy and, 221
Scott, James, 5
Secularization, 27
*Seeing Like a State* (Scott), 5
Sen, Amartya, 5, 220
*Sertão*, 1, 2, 7–8; in Almeida, 125–27; citizenship and, 54–60, *55*; climate of, 16; criticism of development in,

41–43; economy of, 16–21; emergence of term, 69; federal investment in, 97–105; in folk music, 173–76; geography of, 16–21, *17*; health infrastructure in, 48–54; health survey in, 60–65; history of, 16–17; individualism in, 154; in literature, 125–27, 130–31; marginality of, 21–23; modernization and, 21–23, 80; other problems of, 123–24; perceptions of, 15, 24–32, *25*; population of, 180–81, 237n91, 243n3; race and, 22–23
Serviço de Profilaxia Rural (Rural Disease Prevention Service), 46
Serviço Especial de Saúde Pública (Special Public Health Service), 182, 241n27, 246n10
Serviço Geológico e Mineralógico do Brasil (Brazilian Geological and Mineralogical Service), 95
Sharecropping, 25–26, 39, 147
Silva, Severino Borges, 171
Slater, Candace, 41–42
Slavery, 20, 52, 57, 199–200. *See also* Abolition
Small, Horatio L., 95
Smallpox, 51–52
Sociedade Nacional de Agricultura (National Agricultural Society), 68
*Solo e Água no Polígono das Sêcas* (Duque), 156, 157–58
Soper, Fred, 48, 77
Sopper, Ralph H., 95
Souza Brandão, E., 121–22
Soviet Union, 207
Spanish flu, 65–66
Squatters, 19
Stepan, Nancy, 6
Sugar, 18, 193, 194
Superintendência dos Estudos e Obras Contra os Efeitos das Secas (Superintendency of Studies and Works to Combat the Effects of Droughts), 37

Superintendência de Desenvolvimento do Nordeste (SUDENE; Superintendency for Northeast Development), 179, 200–203, 206–11, 216–17

Superstition, 28–29

Tavora, Juarez, 141–42

Taxation, 132, 192, 208

Technocracy, 123–24, 176–77, 178, 220–24

Teixeira, Humberto, 172

Tendler, Judith, 218–19

Tennessee Valley Authority (TVA), 3, 9, 178, 181, 185, 196

Teófilo, Rodolfo, 51–53, 230n24

Tigre, Carlos Bastos, 163

Torres, Alberto, 32, 58–59

Training, agricultural, 162–63

Trinidade, José Augusto, 148–58, 194

"Tropical medicine," 56–57

Trypanosomiase americana, 67–68

Tuberculosis, 49, 69

Tucunduba Dam, 109

Tunisia, 43

Typhus, 190

União Democrática Nacional (National Democratic Union), 205

United States, 99, 100–101, 102–3, 104, 178, 208–11. See also Bureau of Reclamation (U.S.)

Vaccines, 48, 51–53

Vagrancy, 26

Vargas, Getúlio, 11–12, 32, 36, 125, 175; Almeida and, 134–35, 140; Banco do Nordeste do Brasil and, 183–87; engineering and, 88, 140–41; IFOCS and, 127–34; music and, 172, 175; Parente Pessoa and, 196–97; paternalism of, 127–34, 148; revolution, 60

Vidas Secas (Parched Lives) (Ramos), 130–31

Vieira, Luiz, 132, 135–37, 142–43, 142–45

von Martius, Karl Friedrich Philipp, 18

von Spix, Johann, 18

Voting rights, 36, 186–87, 208, 228n44

Vulnerability, 8, 39–40, 123–24, 176

Walker, Gilbert T., 90

Waring, Geraldo A., 95–96

Waste management, 53

Water storage, 91–92, 142–43, 188. See also Irrigation; Reservoirs

Webb, Kempton, 18

Weber, Oswaldo, 91

Whiteness, 22–23, 32, 68, 129. See also Race

Whitening, 57–58, 59, 78, 128

World War I, 98

Worster, Donald, 104

Yellow fever, 51, 66, 76